AF675910

IEE Management of Technology Series 9

Series Editor: G. A. Montgomerie

EXPLOITABLE UK RESEARCH FOR MANUFACTURING INDUSTRY

Other volumes in this series

Volume 1 Technologies and markets
J. J. Verschuur
Volume 2 The business of electronic product development
Fabian Monds
Volume 3 The marketing of technology
C. G. Ryan
Volume 4 Marketing for engineers
J. S. Bayliss
Volume 5 Microcomputers and marketing decisions
L. A. Williams
Volume 6 Management for engineers
D. L. Johnston
Volume 7 Perspectives on project management
R. N. G. Burbridge
Volume 8 Business for Engineers
B. Twiss

EXPLOITABLE UK RESEARCH FOR MANUFACTURING INDUSTRY

G Bryan

Peter Peregrinus Ltd on behalf of the Institution of Electrical Engineers

Published by: Peter Peregrinus Ltd., London, United Kingdom

© 1989: Peter Peregrinus Ltd.

All rights reserved. No part of this publication may be reproduced, stored in a retrieval system or transmitted in any form or by any means—electronic, mechanical, photocopying, recording or otherwise—without the prior written permission of the publisher.

While the author and the publishers believe that the information and guidance given in this work are correct, all parties must rely upon their own skill and judgment when making use of it. Neither the author nor the publishers assume any liability to anyone for any loss or damage caused by any error or omission in the work, whether such error or omission is the result of negligence or any other cause. Any and all such liability is disclaimed.

British Library Cataloguing in Publication Data

Exploitable UK research for manufacturing industry
1. Production management
I. Institution of Electrical Engineers
II. Series
658.5

ISBN 086341 164 9

Printed in England by Short Run Press Ltd., Exeter

Contents

Preface

Every one of the new systems and techniques described in this volume exists because of a real industrial need. Each has been developed and proven through collaborative work between industry and university/college research teams in the UK.

This means that the papers collected here do not only present the latest advanced aids to manufacturing efficiency and control. They also reflect the priorities of the industrial marketplace and contain a valuable database of experience from companies that are advancing the frontiers of knowledge in manufacturing.

The priorities are interesting; it is clear that in general the technology is not the real problem. The hardware largely exists and although there *are* papers here on robots, vision systems and new electrical processes, these are greatly outnumbered by papers on decision aids. Companies have recognised the need to speed up and improve decision-making and have attempted either to capture the knowledge existing in their employees or to optimise their production processes through greater understanding of their key variabilities. Naturally, expert systems feature strongly in the papers.

Managers who are looking forward to modernising their processes by introducing more appropriate manufacturing techniques such as KANBAN, Just-in-Time, Computer-Integrated Manufacturing or an Advanced Manufacturing Technology, will find that a number of the papers address the real problems of planning for new technology. Decision aids are described which include the integration of manufacturing with corporate strategies, the simulation of manufacturing systems and the planning of the other activities which must be rethought to ensure the smooth operation of the investments, such as reward systems, buildings, purchasing and training.

Amongst the technology developments, vision systems are at last seeing real applications. Papers describe the successful use of systems in assembly, welding and food inspection, mostly in conjunction with robots. Robots are no longer portrayed as stand-alone machines, but are shown as integrated sensor-aided systems components which can offer a distinct advantage over other techniques.

A common strand through many of the developments is an appreciation by industry of the need for better quality. The Japanese believe that quality benefits society as a whole and the same could be true of all the developments here if exploited to the full by UK industry.

There can be no doubt about the high quality of the research reported here. It is refreshing to see so much energy and intellect being directed towards solving industry's problems as it approaches the competitive challenges of European (1992 and all that) and global marketplaces. Full credit must be given to the foresight shown by the ACME Directorate of the Science and Engineering Research Council in its timely support of many of the research projects.

The Conference at which the papers collected in this volume were presented was organised by the Institution of Electrical Engineers in association with the Institution of Mechanical Engineers and the ACME Directorate of the SERC.

G C Bryan
January 1989

List of contributors

Professor T. J. Hill
University of Bath

S. H. Chambers
University of Warwick

Dr. K. Chaharbaghi
Imperial College of Science, Technology and Medicine

B. L. Davies
Imperial College of Science, Technology and Medicine

Dr. H. Rahnejat
Kingston Polytechnic

Dr. S. Smith
Sheffield Business School

Dr. D. Tranfield
Sheffield Business School

D. K. Macbeth
Glasgow Business School

L. F. Baxter
Glasgow Business School

N. Ferguson
Glasgow Business School

G. C. Neil
Glasgow Business School

Dr. A. K. Kochhar
University of Bradford

P. Rayner
University of Bradford

G. M. Williams
Plessey Aerospace Ltd

Dr. P. J. Sackett
CIM Institute Cranfield

Dr. J. B. Harris
University of Manchester

Dr. D. J. Walker
Electricity Council Research Centre

Professor B. J. Davies
UMIST

A. T. Joseph
UMIST

M. Kalta
UMIST

Dr. D. G. Elliman
University of Nottingham

C. A. McMahon
University of Bristol

J. H. Sims Williams
University of Bristol

J. Devlukia
The Rover Group

Dr. K. Ridgeway
University of Sheffield

Dr. I. K. Bloor
British Ceramic Research Ltd

Professor S. K. Ghosh
North Staffordshire Polytechnic

Miss T. L. Fitzgerald
University of Liverpool

A. R. Powers
University of Liverpool

J. S. Smith
University of Liverpool

Professor J. Lucas
University of Liverpool

N. Wilmott
Imperial College of Science, Technology and Medicine

Dr. R. Hibberd
Imperial College of Science, Technology and Medicine

Professor W. M. Steen
University of Liverpool

Dr. V. M. Weerasinghe
Imperial College of Science, Technology and Medicine

R. Gibbons
Cambridge University

Dr. D. J. Williams
Cambridge University

D. A. Green
Unilever Research

Dr. A. K. Forrest
Imperial College of Science, Technology and Medicine

D. C. Robinson
Portsmouth Polytechnic

Professor J. Billingsley
Portsmouth Polytechnic

W. J. Finlay
IBM UK Ltd

Dr. J. T. Herd
Heriot-Watt University

N. D. Duffy
Heriot-Watt University

G. P. Philip
Heriot-Watt University

N. J. Eccles
Heriot-Watt University

A. C. Davidson
Heriot-Watt University

Dr. E. K. Lo
University of Salford

A. M. Dore
University of Salford

Dr. P. J. Armstrong
Queen's University of Belfast

Dr. E. J. Wright
Queen's University of Belfast

Chapter 1

Developing a methodology for a business to match manufacturing systems to corporate strategy

Professor T. J. Hill and S. H. Chambers

1 INTRODUCTION

It is now clear that, despite well-publicised improvements in productivity, the UK's share of world trade in manufacturing is still declining. Unprecedented levels of world and local competition, the enhanced expectations of customers and the emergences of new product and manufacturing technologies, are together creating a climate of great turbulence and change, in which only the most competent manufacturing businesses will flourish.

Significantly, however, production decision-making in manufacturing has not changed to meet these new challenges. In most industries manufacturing management still takes a subordinate role in strategic terms to the marketing and finance functions. It still concerns itself primarily with short-term issues (1).

In today's world, businesses must begin to incorporate both manufacturing and marketing perspectives in their strategic decision-making processes. To this end, effective approaches to strategy development must ensure that functional inputs are not only analytical, but also mutually understood and agreed.

2 MANUFACTURING STRATEGY

Writers in the U.S. such as Hayes and Wheelwright(2) and Skinner(3) have emphasised that a business can use manufacturing as a powerful strategic asset. However they propose that, in order for this to be achieved, the company must be able to define a particular task (or tasks) for which manufacturing must then be appropriately equipped and organised. Such tasks include low cost manufacturing, high quality, or delivery performance. Evidence is provided that some businesses that have achieved these objectives are highly successful. But, little guidance is given to illustrate how these tasks should be selected as the basis of a manufacturing strategy. There is an implied assumption that the executives should know what is required. In practice companies rarely provide a clear indication of what single competitive edge should be provided by manufacturing. Normally, it is required to react to what is asked of it and must cope with resolving the conflicting demands which seem appropriate at the time. As a result, performance in any single task is compromised, and manufacturing remains vulnerable to criticism for its inadequate achievements. The consequences extend beyond inter-departmental conflict; the business suffers because it fails to provide the customers' needs. In addition, all too often, what appear to be routine manufacturing decisions frequently come to limit the company's strategic options, binding it with facilities, equipment, personnel, and basic controls and policies to a non-competetive posture which may take years to turn round(4).

Manufacturing's role in a competitive strategy can be of two generic types. The first is to provide a unique and protected manufacturing process technology which offers a sustainable marketing edge. Examples of this are relatively rare since most process technology is available on the open market or can be identified and copied. The second is for manufacturing to provide co-ordinated support for the ways in which a company wins orders in the marketplace, at a level which is better than the manufacturing functions of its competitors are able to do. In order to achieve this, manufacturing must develop congruent policies for both its processes and its infrastructure (controls, procedures, systems and structures), which support the market requirements. A framework proposed by Hill(5) outlines the basic elements that can be considered in this process, as shown in Table 2.1

TABLE 2.1
Framework for reflecting manufacturing policy issues in corporate decisions (steps involved)

1	2	3	4	5
Corporate objectives	**Marketing strategy**	**How do products win orders in the market place**	**Manufacturing strategy**	
			process choice	**infrastructure**
• growth • profit • return on investment • other financial measures	• product markets and segments • range • mix • volumes • standardisation versus customisation • level of innovation • leader versus follower alternatives	• price • quality • delivery speed reliability • colour range • product range • design leadership	• choice of alternative processes • trade-offs embodied in the process choice • role of inventory in the process configuration	• function support • manufacturing systems • controls and procedures • work structuring • organisational structure

Unlike previous approaches, this framework directs executives to consider the essential links between marketing, the market, and manufacturing through an assessment and understanding of its order-winning criteria.

3 THE RESEARCH

The objective of the ACME-funded research was to test and refine the framework/methodology in a number of companies in diverse industries and with a variety of process choices. In each case the level of consistency was to be examined and possible changes identified. The research objectives were to

a) observe and record the characteristics of any existing approaches to relevant strategy development, with particular emphasis on their impact on manufacturing.

b) refine the framework and methodology so that it could be applied in a wide variety of situations.

c) improve the conceptual and language base available for manufacturing strategy development.

Field research was carried out between range October 1985 and February 1988 in eleven companies in a range of industries. In many it was found appropriate to concentrate on only those parts which were identified to be of particular importance to the business, and this allowed greater detail of specific aspects of manufacturing strategy developments to be considered. The result was the opportunity to refine existing and to develop additional concepts and approaches.

This paper does not include all the detailed developments that have come out of the work. Rather, it is intended to provide insights into the nature of the refined approach, and the advantages this offers compared to traditional methods of developing strategies. Also, the paper will select and overview some of the important developments in concepts which have practical application in the appraisal and design of manufacturing systems best alligned to the needs of the market.

4 EXISTING APPROACHES TO STRATEGY FORMULATION

In most companies reviewed, formal strategy development had been undertaken during the last few years. In each instance, the obvious complexities had been addressed by splitting the task into its functional parts. Marketing strategies generally took account of the opportunities provided by fully or partially completed product developments, and for this reason, the views of design/ development executives had been taken into account. In other respects, marketing strategies had been developed in relative isolation from the rest of the business.

The credibility of marketing approaches had been typically enhanced by analysis of market segments and market share, and by reviews of competitors' strengths and weaknesses. Since this work is often beyond the resources of the marketing department, it was sometimes contracted out to external consultants or market research firms.

This approach further detaches the marketing policy decisions from the realities that will be faced by manufacturing. Marketing descriptions and averages frequently obscure significant market differences which will ultimately affect the competitiveness and profitability of the business.

Other functional strategies and particularly manufacturing's (where they exist) follow a similar pattern, but with few standard approaches or concepts available, often comprise statements listing means of achieving "improvements" and setting appropriate targets. Invariably, this leads to very comprehensive lists of actions and anticipated results, but rarely is any basis given for checking the relevance of, or prioritising this work.

As with marketing, manufacturing strategies are made more creditable to the executive team by the inclusion of reports by technical specialists. However, the manufacturing function always failed to provide the strategic parameters in which these contributions should be placed. Typically, the result was that manufacturing strategies became dominated by technical perspectives in much the same way as marketing reports are biased towards their specialists' view of what is best for the company.

Since a business strategy comprises the composite of the individual submissions, there is no effective mechanism to ensure that the critical issues are seriously considered and debated by every function to ensure external consistency with the needs of the markets and internal consistency between the parts of the business within that provision. The research has shown that in each company, long-term decisions had been or were proposed to be taken that failed to provide this match, with actual or potential damage to the competitiveness and/or profitability of the company.

However the greatest concern, was that many senior executives, despite their extensive knowledge of the company, its products, and its function, lacked a detailed understanding of the business attributes of the company they controlled. This was highlighted in many ways, for instance:

a) when they tried to understand and explain why major investment (in processes or infrastructure) had subsequently not provided the anticipated benefits or had created unexpected disadvantages. For example, a packaging manufacturer found, contrary to expectations, that profitability fell and delivery speed provision deteriorated after their investment in a multi-million pound "state of the art" printing press. A small aerospace components manufacturer invested nearly £1m in a computerised planning and control system without achieving any of the anticipated reductions in delivery lead times, inventory levels and costs.

b) when they were initially unable (and sometimes unwilling) to clearly analyse the business implications of the orders they receive. A car components manufacturer had promoted increased sales of its most popular products without knowing that these items provided almost no contribution, in contrast to its many high margin low-volume products. An injection moulding business, dividing its manufacturing into a number of self-contained plants, failed to recognise the opportunities for different infrastructure in each, in line with the differing market characteristics of its products.

5 METHODOLOGY

5.1 Initiating The Work

The framework requires that the essential link between marketing and manufacturing is forged by analysing the order winning criteria of a company's products. As executives are unused to taking this approach, any manufacturing strategy development must begin with an introduction to the basic concept of the approach, and an explanation of the need to adopt standard terminology. When executives are asked to define order-winning criteria, there is often a reluctance for individuals to commit themselves, so it can often be helpful to use small groups of appropriate individuals (e.g marketing directors and sales managers and representatives). The group involvement encourages debate and resolution of the problem. It is, however, critical to provide some mechanism for challenging loose descriptions. For example, people often refer to "service" as an order winner, but this is too vague to be of use: they must be encouraged to be more specific (e.g. delivery speed, design support, salesperson's skill).

5.2 Use Of Representative Samples For Market Analysis

In most manufacturing businesses there is a wide variety of different markets, customers, products, and types of order, and the mix of these factors is changing over time. Some participants may suggest that the resulting complexity is such that it is futile to attempt to define order winning criteria for any parts of the whole. Hill(4) proposes that marketing should review all current and proposed products and to divide them into types having similar order-winning characteristics. Where there is a large number of products, representative product samples need to be used.

The rationale for the choice of representative samples will be different from company to company, and will rely on the judgement of the executives involved. They must be encouraged to select the initial sample so that the most important differences are recognised. Factors which may be considered include market segments, product types, customers, major processes used, volumes, and order characteristics.

It is critical that the chosen sample is finally

acknowledged as representative by the executive team. The sample may be added to as the work proceeds, as new important differences are identified.

The research has shown that marketing descriptions alone (segments, customer groupings and product types) are not a suitable basis for manufacturing strategy development, as they conceal critical differences with which manufacturing must address. For example:

Steel Sections Manufacturer
"Customised Sections" market was shown to comprise four significantly different manufacturing categories (special or standard steel, and delivery speed or normal lead time)

Automotive Electrical Manufacturer
"Alternators - UK Agents" market consisted of extremely wide spread of volumes, with order-winners ranging from predominantly price to entirely delivery speed.

Carton Manufacturer
"Beverage Cartons" market included a number of customers requiring exceptional levels of quality and delivery speed but this work gave high margins. Marketing initiatives brought in major new "beverage" business which was extremely high volume and price-sensitive.

5.3 Defining Order-Winning And Qualifying Criteria

The next step is to allocate weightings to the groups of products having similar order winning and qualifying criteria, for both the present and two time periods in the future. The research has shown that this is best done initially by marketing and sales personnel who should have the greatest understanding of the markets and contact with customers. This step should not be constrained by any prescribed check list, and should recognise any seasonality or product life cycle effects. The allocation of weightings involves awarding percentages to each order-winning criterion, which reflects its degree of importance. The criteria may be manufacturing-related (e.g. price, delivery speed) or non-manufacturing (e.g design support, sales representatives' skills). In a plastic products business 'technical competence' was identified as significant in several products, and this was related to a combination of processing advantages, matched with appropriate product designs/specifications.

Differences between the characteristics of repeat business and new or potential business must be recognised in this analysis. For example, in several of the companies new orders were seen to be significantly more price-sensitive than existing work of a similar type.

When a complete list of order winning criteria has been prepared, it is then offered to the other functions for comments. Frequently, the manufacturing perspectives are quite different to those of the sales/marketing functions. This seems to be because manufacturing must cope with the demands of every individual order whereas sales contact normally orientates around the longer term requirements. By identifying differing perspectives, inter-functional debate is created and common business language (order-winning criteria) is used for the first time. This has two consequences: it enables important business issues to be brought to the surface, and it provides the basis for emphasis in further analysis work.

5.4 Further Analysis Based On Order Winning Criteria

The marketing and manufacturing assessment of order winners enables subsequent analysis work to be directed towards those factors that are important for the particular business. For example, where price is seen to be a dominant order-winner for a group of customers, products, or orders, an analysis of costs and selling prices would be undertaken for each part of the representative sample. Where delivery speed is a factor, a summary of actual demands for reduced lead-times and schedule changes is prepared.
The outcome of these analyses includes:

a) identifying differences between customers, products and orders which are sometimes included in the same initial order-winning group. Where this happens, additional order-winning groups are created.
b) evaluating the extent of relevant order-winners in terms of (for example) costs/price, short lead times, quality levels and delivery reliability.
c) to identify the extent to which manufacturing fails to meet b), for example
    - actual costs higher than estimated
    - inadequate delivery reliability
    - quality complaints/returns
d) confirm or modify functional views of order-winners and their weightings

5.5 Advantages of the Methodology

Many benefits of the approach have been revealed by the research. Firstly it allows a business to gain an increased understanding of its markets over time, as the interchange of analyses and debates takes place. This is revealed by:

- declared changes over time in the assessed weightings of order winning criteria
- changes in the perceptions of, and attitudes towards markets and customers
- marketing's "unsuitable" segments being shown to be very suitable for the company overall
- markets originally perceived as only of average importance being re-analysed into highly suitable and unsuitable segments

The approach avoids the use of checklists or routine

analysis in strategy developments and enhances the executives' awareness of their ultimate responsiblity for business strategic development.

It ensures that customer needs are given greater consideration than marketing dogma or convenience.

Finally, it encourages companies to consider that its manufacturing processes and infrastracture can be developed as strategic assets without having to turn to externally-derived 'solutions' to manufacturing's problems.

6. MANUFACTURING STRATEGY CONCEPTS

Our research has provided the basis for several important conceptual developments which can be used in manufacturing strategy development. Two, of general interest, are summarised below

6.1 Focus

The concept of focused manufacturing was first proposed by Wick Skinner,(6). He highlighted the fact that most existing plants are expected to produce many different products, for numerous customers, in a variety of markets with the corresponding multiplicity of conflicting manufacturing tasks. This complexity leads to many compromises, to high overheads, and to numerous situations where manufacturing is open to widespread criticism. By limiting each plant to a "concise, manageable set of products, technologies, volumes and markets", the process and infrastructure elements of manufacturing can be "focused" and the resulting performance should be enhanced: simplicity, repetition, experience, and homogeneity of tasks breed competence. Skinner claimed that the focused factory will outproduce, undersell, and gain competitive advantage over the complex factory.

The implication of this concept is that a company should consider either limiting the complexity of its product range, markets, volumes and customer requirements; or dividing manufacturing into specialist "plant-within plant" configurations. For many, the second has greatest appeal (as it does not require short-term reductions in sales), but the basis of the changes must be carefully considered, as there are always many options, each with its own set of trade-offs.

The division of manufacturing into smaller relatively self-contained units can be achieved in different ways.

a) By product name or market segment.

This approach relies on marketing descriptions of products and markets, and provides an orientation towards customer needs and priorities. However, such plants would still be expected to cope with a range of volumes (e.g. products at different stages of their life cycles) and order winners (e.g. som dominantly won on price, others on delivery speed). The plant may be smaller, but the breadth of complexity would be the same.

b) By process.

In most business, products follow different process routes. It is often possible to identify families of products which have similar processing requirements, and to establish self-contained plant-within-plants equipped to produce them (known as Group Technology). With this approach, as with product/market divisions, the range of manufacturing tasks and volumes will normally be wide and relevant order-winners, disparate.

c) By volume

An approach which divides manufacturing on the basis of manufacturing volume provides the opportunity to maximise the benefits of appropriate process choice. Thus, high volume products are made in line or batch processes with extensive dedication, and very low-volume products are produced in a flexible low volume batch or jobbing plant. As products pass through the various stages of their life cycle, they must be transferred from plant to plant.

In some businesses, order winning criteria may be related to volume, but in many cases that is not so. It is possible, therefore, for such plants to have very mixed manufacturing tasks, with the inherent complexity remaining unresolved.

d) Focus

Two of the characteristics of the focused factory as described by Skinner, were that it should be required to do a "superb job on one or two market demands" and that it should "excel in one manufacturing task". The research has shown that these aims can be fulfilled by choosing a group of products on the basis of similar order-winning criteria. Normally this will involve some duplication of plant, but there are unique benefits in the design of infrastructure. Since each plant-within-plant is focused on an explicit manufacturing task, much of the manufacturing infrastructure can be built-in exactly as required, eliminating much of the need for centralised departments. Thus, a plant-within-plant specialising in non-standard "quality" will be able to specify its precise requirements for inspection, (personnel and procedures), whilst other plants may adopt simpler and cheaper solutions for their needs. A plant specialising in "price" will ensure low costs by different planning approaches, order quantity sizes, shift arrangements, and work structuring, than one providing "delivery speed".

The research has shown that many companies recognise the benefits derived from dividing manufacturing into more self-contained and autonomous units. However, the approach to the division is often orientated to marketing descriptions or engineering solutions, rather than through an understanding of the manufacturing tasks involved i.e. manufacturing strategy.

6.2 Product Profiling

A company needs to have a comprehensive understanding of the changing implications to its business as alternative processes are chosen, and then to use this as an important input into the corporate debate. It must be able to evaluate the degree of fit between its manufacturing strategy and the order-winning criteria of its selected markets. Product profiling (7) is one concept that can be used in this assessment.

The research has enabled this approach to be developed to take account of many products being manufactured on several processes. It has enabled companies to understand their strategic options, to select appropriate processes, and to pursue markets which manufacturing is best able to support.

REFERENCES

1. Hill, T.J., 1985, 'Manufacturing Strategy', MacMillan Education Ltd, England.

2. Hayes, R.W. and Wheelwright, S.C., 'Restoring our Competitive Edge - Competing through Manufacturing', 1984, John Wiley, USA.

3. Skinner, W., 'Manufacturing, the Formidable Competitive Weapon', 1985, John Wiley, USA.

4. Skinner, W., 'Manufacturing - Missing Link in Corporate Strategy', Harvard Business Review, May-June 1969, p 136.

5. Hill, T.J., 1985, 'Manufacturing Strategy', p 41.

6. Skinner, W., 'The Focused Factory', Harvard Business Review, May-June 1974, pp 113-21.

7. Hill, T.J., 'Manufacturing Strategy', pp 89-95

Chapter 2

Intelligent policy-making for computer-integrated manufacturing systems

K. Chaharbaghi, B. L. Davies and H. Rahnejat

ABSTRACT

This paper presents the outcome of a detailed research study which has been supported by the UK Science and Engineering Research Council under the ACME initiative. This study has resulted in an advanced operational control methodology which generates sound, accurate and trustworthy manufacturing strategies by incorporating a goal seeking mechanism within a discrete-change simulation medium. The methodology makes use of performance indicators which are predicted by simulation in order to determine whether the specified goals are satisfied. Failing to meet the goals, the manufacturing system is redefined accordingly through an iterative process until the simulation predicts that the desired goals are achieved, thus culminating in the determination of an optimal policy. This approach has been successfully applied for obtaining manufacturing launch date strategies in multi-variant-multi-batch production environments.

1. INTRODUCTION

Computer-integrated manufacturing (CIM) encompasses a number of activities that include CAD/CAM, CAPP/CAPM, automated tooling and material handling, process monitoring and fault diagnostics amongst a host of others [Boaden and Dale(1), Ranky(2), Sata(3), Leonard(4) and Groover(5)]. CIM is a generic concept which incorporates the computer control of all managerial as well as technical functions related to progress of manufacture from receipt of customer orders to dispatch of finished items. CIM promises to offer a dramatic reduction in manufacturing lead times and give a radical improvement in the level of inventory and exploitation of resources. The key to this, however, hinges on the definition of a well-defined manufacturing control methodology.

A number of CIM systems exist both in the UK and elsewhere where specific manufacturing control solutions have been found to suit their particular requirements. Very little research has been directed to identifying system parameters under various manufacturing scenarios and formulating a general methodology that apply these to enable optimal, or near optimal, control for the progress of manufacture. Buzacott (6), Dupont-Gatelmand(7) and Buzacott and Yao(8) have carried out a number of surveys of the manufacturing control methodologies currently employed. These have revealed a number of differences in terms of formulation and application. There is a major need for research to narrow the large gap that exists between the problems inherent in the formulation of current control methodologies and the problems associated with controlling CIM systems.

The accomplishment of a coherent control for CIM systems requires sound strategies that enable the achievement of production requirements, minimisation of inventory levels and effective exploitation of manufacturing resources. The generation of these strategies calls for the provision of a vast amount of production data which must then be processed by a computer-based intelligent tool. To achieve this in real-time should be the ultimate objective but this is a tall order given the current state of the art. However, an intelligent system off-line, which will formulate the appropriate strategies to be adopted for on-line control of manufacture, is quite feasible.

The research work herein presented has aimed to form such an intelligent methodology. The formulation of this methodology has been intended to be general in order to have a wide range of applicability.

2. OPERATIONAL CONTROL OF COMPUTER-INTEGRATED MANUFACTURING SYSTEMS

In the traditional manufacturing scene, the interfaces between crucial elements of the system for operational control are human controlled. The collection and feeding of information to and fro, between various activities, are performed by the personnel such as foreman, planners, controllers and management. However, they have proved to be the weak link in the process of manufacturing control. There are many reasons for this but those chiefly cited include the tiresome monotonous collection of data, oversight in detection and reporting of up-to-date information, untimely data gathering and refusal to cooperate. It is therefore not surprising that planners have sought to introduce more and more automation in order to curb the role of personnel in the manufacturing control.

Since the introduction of numerical control technology little has been gained from the manufacturing control viewpoint because reductions in processing times and set up durations have not been translated into a corresponding reduction in inventory levels. Until recently it was perceived that the blame for all manufacturing control problems and production inefficiency should be put squarely at the doorstep of a large direct labour content of jobs. There is a strong feeling in industry that this may no longer be true. In fact most of the blame can be now attributed to slack production control and inadequate means of meeting the rapidly changing demands using policies based on traditional wisdom. This situation is only reversed when innovative operational control methodologies are devised and implemented. Therefore, whilst CIM requires a sound integration discipline, it is in dire need of intelligent operational control interfaces that require methodological implementation rather than collection of large data streams and their manipulation by rules of thumb. CIM will be a reality when such intelligent controls are developed and employed.

The problems of manufacturing control are in many ways combinatorial. There are a large number of interdependent parameters such as work-in-progress inventory and resource utilisation which are desired to be kept at certain levels but are often conflicting in nature. The fascination with the combinatorial nature of manufacturing control problems such as scheduling has led to the development of an assortment of analytic methods and simulation studies, largely of an academic nature (particularly with the former).

The usefulness of these studies has been limited primarily for two reasons. Firstly, the means required to implement them has not always been available. For instance, the acquisition of real data in real environments and in a reliable form has always been an obstacle due to the human related factors mentioned earlier. Secondly, complex and cumbersome algorithms have required data processing and analysis cycles that do not comply with the need for providing quick answers to an immediate and pressing problem. The complex nature of multi-parameter manufacturing control has also resulted in the creation of complicated methods that can be applied to rather limited problems. This is because their modelling framework makes unrealistic assumptions and there are many computational restrictions. Therefore the question arises: where does one go from here and can CIM become a reality?

Hitherto, it has been revealed that the effective control of manufacturing activities can be attained by defining meaningful key parameters of manufacturing effectiveness. The selection of such parameters is difficult, not withstanding the obvious, such as the resource utilisation rates, throughput times, production rates, etc. Obviously, the relevance of these parameters to given circumstances assumes the key importance in their use. As is often quoted in academic exercises, one may use a host of these parameters and optimise the performance of a fictitious system with respect to some or all of them. But this is an academic prerogative and is not generally an industrial concern.

Industrialists may well question the need to entangle themselves with laborious mathematical scheduling and production control algorithms that still fail to solve such basic problems as job-shop scheduling using a few unreliable resources and half a dozen jobs of various types. This is the reason why the technique of simulation has been increasingly applied since it is based on modelling the systems by interpretation rather than by analytic conceptualisation. In simulation, rules, facts and relationships are used to represent real systems by computational models that are based on an assumption-free framework. A lot of literature exists on the technique of simulation, mostly about the various simulation moedlling methods and its application in configuring systems through scenario-building processes [Chaharbaghi(9), Chaharbaghi and Davies(10), (11), and Chaharbaghi et al (12), (13)].

But how can the simulation technique help us with the development of methodologies that are required to provide the intelligent integrative interfaces for CIM systems? The answer is in the off-line implementation of proposed methodologies, without the need for real-time testing in the development stages.

The pivotal aim of CIM is to meet the customer demands on time and as cheaply as possible, whilst maintaining an appropriate level of resource exploitation and an acceptable quality. The manufacturing costs may be squeezed by automation, which is not a major manufacturing control concern. However costs may also be squeezed by reducing the capital cost tied-up in the form of stock, which is of a prime concern to production controllers.

In the next section of this paper it is shown how relevantly defined manufacturing control parameters can be employed to enable policy generation through goal seeking. The appeal of the methodologies presented in this paper lies in their simplicity, relevance and effectiveness.

3. INTELLIGENT POLICY-MAKING FOR CONTROL OF CIM SYSTEMS

Having established the need for a methodological policy-making process that is based on simulation studies, the next step is to define the way in which this can be developed and implemented. Hitherto, a number of simulation methods have been proposed to solve production control problems. The most common approach is based on the definition of rules and selection and implementation of these within a scenario building simulation exercise. The shortcomings of this approach are often the large selection of rules that can be logically employed and the combinatorial trial and error application of these in the simulation study. This open-loop approach to simulation has always limited its use to simple manufacturing control problems.

The above shortcomings can be overcome by providing the simulation with a closed-loop approach, incorporating some degree of intrinsic intelligence within its modelling environment. A general example of such methodology is given in figure 1. As shown in this figure, the simulation modelling environment is provided with a mechanism which monitors the system performance, compares it with that predicted by the simulation model, and thus ascertains the degree of conformity with a predefined goal. This process forms the criterion that provides the diagnosis of the system state with a given policy. If the system state meets the intentions of the current policy, the goal is deemed as realised, otherwise a prognostic action has to be undertaken. This action readjusts the controllable parameters of the system thus forming the new policy. This procedure is repeated until the intended goals are attained.

One of the example areas where the above generalised methodological policy making concept can be applied is the generation of launch date strategies in the operational control of manufacturing systems. These strategies highlight when the raw material should be launched to the manufacturing systems so that the required goods are dispatched to the store on the required date. In some companies there is no strategy, as such, and the raw material is launched as soon as a job order is placed by the customer. The danger with such an approach is that the jobs are completed either too early or too late depending on how the system is loaded. The former results in excessive inventory cost in the form of finished goods while the latter leads to high levels of work-in-progress inventory as well as customer dissatisfaction. In other companies, the precalculated job manufacturing cycle times and due-dates form the basis of the launch date strategy. To compute the job manufacturing cycle times, the processing times are added together and an average queuing time is also accounted for. The job manufacturing cycle times are then deducted from their respective due-dates to form the launch dates. The application of this launch date strategy will also have some adverse effects. This is because when the system is lightly loaded, the actual queue times will be less than the allowed ones so that jobs are completed ahead of their due-dates, thus resulting in excessive levels of finished goods inventory costs. Alternatively, when the system is heavily loaded, the actual queue times will by far exceed the allowed average queue times, thus resulting in substantial lateness in meeting the specified due-dates.

Figure 2 gives the procedure which, when employed, streamlines the current launch date strategies and hence establishes the appropriate ones. This procedure first simulates the manufacturing system under study using the current launch date policy. If this simulation indicates that with such launch dates all the jobs are manufactured just-in-time, the goal is considered as achieved and the launch dates are deemed appropriate. Otherwise the launch dates are adjusted accordingly. This procedure is repeated in an iterative manner until the simulation predicts that all the jobs are manufactured just-in-time. Here, it must be pointed out that the methodology in the form presented in Figure 2 has one limitation and that is there could be a number of additional constraints that may impede the achievement of the desired launch dates. One common constraint is that some jobs are required to be launched beyond the beginning of the investigation. The inadequacy of system design can be another cause of such an impediment. For example, some resources can be overloaded during the entire period of investigation. However, this problem can be overcome via a tuning process. For instance, overloaded resources can be relieved by deploying redundant resources, extending the production shifts through overtime, etc.

The procedure herein described has been applied to a real industrial system sited at the Leavesden plant of Rolls-Royce plc. The results obtained are summarised in figures 3 and 4. Figure 3 compares the repercussions resulting from the traditional launch date policy to that generated by the policy maker after six iterations for 12 job orders. A positive value in this Figure indicates lateness in meeting the specified due-dates while a negative value implies the reverse. Figure 3 justifies the adoption of launch dates generated by the policy maker as the outcome is much closer to the concept of just-in-time manufacturing. Figure 4 compares the utilisation and manufacturing resources stemming from the traditional and policy-maker launch date configurations. This figure reveals that resource utilisations for both configurations are very similar. The reason for this similarity is because the production load for both configurations has remained unchanged over the period of investigation. From the above comparisons it can be concluded that the policy-maker approach results in a more balanced launch date configuration. It should be pointed out that the methodologies presented in this paper are in their simplest form. Additional features can be incorporated in these methodologies to enhance the generated policies. For example the effect of perturbations (such as machine breakdown) on the manufacturing lead times can be ascertained and taken into account in determining the launch dates.

4. CONCLUSIONS

This paper has highlighted the need for a methodological approach to policy-making processes. It has shown that such an approach has to incorporate an element of intelligence within a simulation environment to simplify the combinatorial type problems into one simple diagnostic and prognostic procedure. The elements of simplicity, prediction, diagnosis and prognosis form the basis of an intelligent policy appraisal process. One such approach has been used successfully for the determination of launch dates in a multi-variant-multi-batch production environment. The outcome of the exercise has demonstrated that a large class of manufacturing control problems with excessive inventory levels can be tackled and yield a just-in- time production methodology. In these circumstances the original aims of reducing the inventory levels whilst meeting the specified due-dates have been realised. The methodology presented here can just as easily be employed for other applications which require a process of policy making based on intelligent interpretation of simulation predictions in rapidly changing dynamic environments.

REFERENCES

1. Boaden, R.J., and Dale, B.G., 1986, 'What is Computer Integrated Manufacturing?', Int. J. Operations and Production Management, 6, 30-37.

2. Ranky, P.G., 1986, 'Computer Integrated Manufacturing: An Introduction with Case Studies', Prentice-Hall.

3. Sata, T., 1987, 'Recent Developments in Computer Integrated Manufacturing in Japan', Robotics and Computer Integrated Manufacturing, 3, 373-380.

4. Leonard, R., 1988, 'Elements of Cost-effective CIM', Int. J. Computer Integrated Manufacturing, 1, 13-20.

5. Groover, M.P., 1987, 'Automation, Production Systems and Computer Integrated Manufacturing', Prentice-Hall.

6. Buzacott, J.A., 1985, 'Modelling Manufacturing System', Robotics and Computer Integrated Manufacturing, 2, 25-32.

7. Dupont-Gatelmand, C., 1983, 'A Survey of Analytical and Simulation Models for the Design and Control of Flexible Manufacturing Systems', Manufacturing Systems, 12, 157-164.

8. Buzacott, J.A., and Yao, D.D., 1986, 'Flexible Manufacturing Systems: A Review of Analytical Models', Mananagement Science, 32, 890-905.

9. Chaharbaghi, K., 1987, ' A Computer Simulation Methodology for planning the Design and Operational Philosophy of Advanced Manufacturing Systems', PhD Thesis, Imperial College, University of London.

10. Chaharbaghi, K., and Davies, B.L., 1986, 'Manufacturing Sytems Simulation using DSSL', CAD, 18, 242-256.

11. Chaharbaghi, K., and Davies, B.L., 1986, 'The Analysis of Flexible Manufacturing Systems for Reliability', Int. J. AMT, 1, 79-100.

12. Chaharbaghi, K., Davies, B.L., and Rahnejat, H., 1986, 'Application of Simulation in Planning the Design of Flexible Assembly Systems', Proc. 4th Conferences on UK Research in Automated Manufacturing, London, England.

13. Chaharbaghi, K., Goddard, A.J.H., Sayles, R.S., Bunde, R., and Khodabandehloo, K., 1988, 'Performance Simulation of Next European Torus', Proc. 6th International Conference on Reliability and Maintainability, Strasbourg, France.

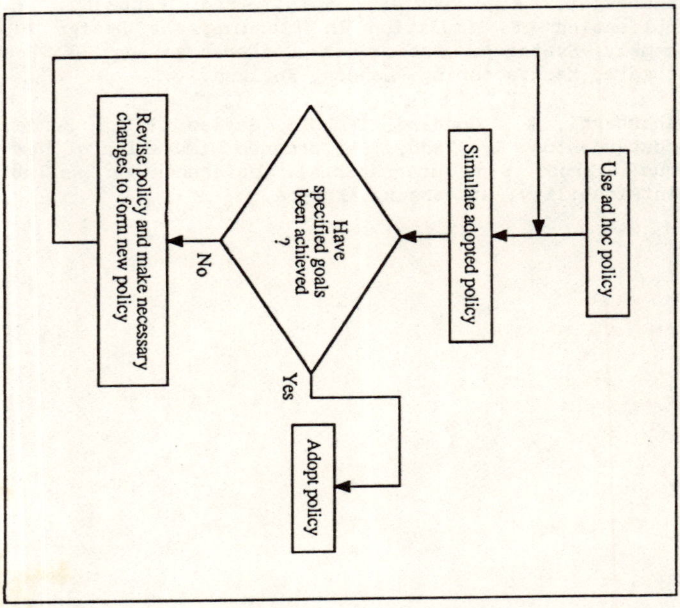

Fig. 1. A methodological procedure for policy-making

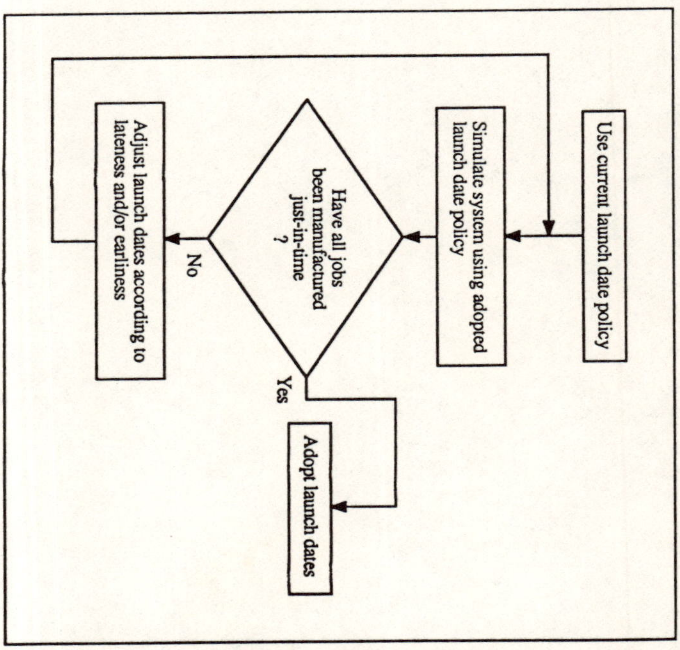

Fig. 2. Procedure employed for determination of launch dates

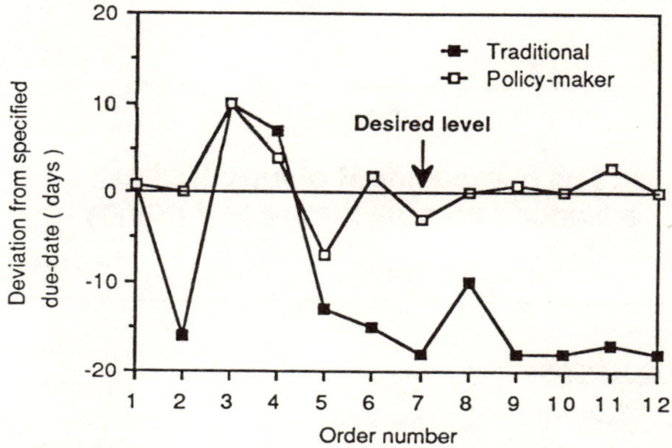

Fig. 3. Comparison of success in meeting due-dates for differing strategies

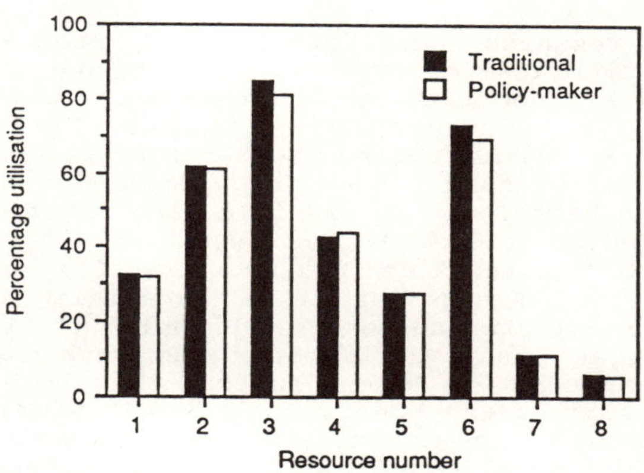

Fig. 4. Comparison of resource utilisations for differing strategies

Chapter 3

The management of introducing advanced manufacturing technology

S. Smith and D. Tranfield

1. INTRODUCTION

"Implementation is the process of taking a technically sound complete system and making it operate effectively in the real environment of the business world". (3)

From our observations of a limited number of U.K. manufacturing companies there is often a failure to distinguish between installation and implementation in the introduction of Advanced Manufacturing Technologies. All too often it seems that those charged with introducing these systems use narrow technologically dominated methodologies which address installation and fail to lead to implementation. Fortunately also we did find some companies who in our opinion were very successful in implementing A.M.T.

Other research and reports of companies and consultants practical experiences suggest that our limited observations are not isolated and that many companies find difficulty in successfully implementing computer applications in manufacturing whether the applications are in Computer Aided Manufacture (e.g. Computer Numerically Controlled Machine Tools to Flexible Manufacturing Systems (2,4)), Computer Aided Design (e.g. Computer Assisted Draughting to Computer Aided Engineering (12)), or Computer Aided Production Management (e.g. Materials Requirement Planning to Manufacturing Resource Planning (6, 13)). Most Findings suggest that the greater the move to Computer Integrated Manufacture the greater the implementation problems. Waterlow and Monniot (13) and Meredith and Hill (6) have proposed similar conceptual schema for classifying the level of integration of manufacturing systems and have identified distinct implementation problems associated with each level. Despite this plethora of research, relatively little work has been addressed at identifying the elements of approaches which lead to successful implementation of Advanced Manufacturing Technology. This research set out to make a practical contribution to this area by examining a number of A.M.T. applications in depth and attempting to identify which situations and approaches/methodologies to implementation were considered to lead to successful outcomes and which were not. From this, we distilled the main features necessary for a successful implementation methodology.

2. APPROACH

From the outset of this research it was apparent that A.M.T. implementation needed to be studied in the broader context of the business and organisation within which it was to be exploited. It was also clear that the implementation process took time and that the "history" of the implementation needed to be traced, identifying the key managerial actions and the changing nature of the external competitive environment. For these reasons an in depth case study approach was adopted. Although this limited the number of applications examined (3 examples of CAPM, 2 CNC, 2 CAD/CAM, 3 CMS, 2 Flexible handling/assembly systems), it did allow the richness of data necessary to begin to understand the complexities and contingencies of implementation processes adopted.

3. FINDINGS

Many of the generalised findings have been reported elsewhere (7,8,9,11). Fig i. below contrasts two archetypical examples drawn from real cases of contrasting approaches implementation in manufacturing. These contain, in our opinion, the main elements of successful and unsuccessful attempts to implement AMT.

Fig.i. Implementation Details of Two Archetypical Companies

	Company X	Company Y
Application	To automate exist ting operations	To secure strategic advantage
Project Management	Steering group/Small Project team	Involvement throughout management
Outside Consultants	Strictly technical input - suppliers	Used to complement in-house capability
Management	Top management only	All management
Training Focus	Technical aspects only - limited general education	Wide general education - workshops on local
Training Spend	2% Annual labour cost	10% Annual labour
Reorganisation	Job Design only	Company Reorganisation
Reward Systems	Not considered until late in the implementation	Grades reduced from 17 to 5 - single status employer
Associated	None planned at out-set - some consideration of CAPP later	Total manufacturing systems review - JIT, TQM etc implemented as part of change
Time Scales/Focus	Related to installation of technology	Related to business goal
Outcomes	Union Dispute Poor Planning Late Implementation Limited impact	Implemented early Slight underspend Massive productivity Manpower down factor 3/4 Supplier base down factor 4

4. CHARACTERISTICS OF SUCCESSFUL IMPLEMENTATION

Although there are obvious dissimilarities in terms of scale of investment, type of technology, the type of environment etc. within the two examples above, our experience was that both typified two fundamentally different approaches to AMT implementation. From our research, the more successful companies could be distinguished by a number of features of their implementation strategies:

. As a first step, they conducted a severe critical appraisal of the company in business terms and built a vision of business objectives in which the role of manufacturing was clearly defined.

. They combined a technological configuration and organisation design to comprise a delivery system to achieve the pre-stated ends, thus initiating change on a wide variety of fronts.

. They used outside specialists in a complimentary way to achieve strictly defined company objectives.

. They clearly signalled to the management team and the workforce the new strategic direction and role of the AMT. Particularly the refocussing of reward and control systems and the use of training and development strategies were important to achieve this.

. They initiated the process top-down but ensured that the implementation process gave ample opportunity for bottom-up involvement to build commitment to the new technology and working practices.

. They specified a clear time focus for the initial change effort and a process of continuous improvement thereafter.

All of these features we attempted to building into the generic change methodology which we developed.

5. OUTLINE STRUCTURE AT IMPLEMENTATION METHODOLOGY

In developing the outline of a structure for a methodology, we took three main factors into account:

(i) it should include the most important features reported in the methodologies adopted by the Type Y (see Fig.i) companies studied

(ii) it should incorporate a means of achieving integrated step change

(iii) it should address a wide range of managerial, technological and organisational dimensions.

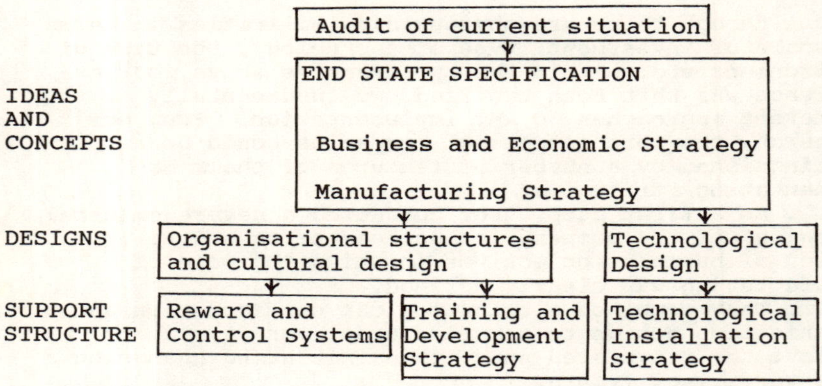

Fig.ii An outline of the implementation methodology.

6. BASIC COMPONENTS OF THE METHODOLOGY

Whilst the methodology briefly outlined above provides a roadmap for the implementation of AMT, it does not identify the key managerial activities involved, nor does it mention time scales and the pace of change. From our research we have been able to identify the following four key managerial activities in implementing AMT:

(i) Audit and Vision Build - this has been outlined above and is a top management activity which involves looking outward from the company and identifying market standing on key competitive dimensions. The main aim is to produce a business plan complete accompanied by a set of central values (no more than 4-6) with which company members can identify.

(ii) The Cascade - once the initial work has been done and agreed by the top management, then the "message" (both business plan and central values) can be cascaded through the management team. This may be done after structural reorganisation, and will usually involve the use of training and development as well as change in reward and control systems. A cascade is usually best handled by a series of workshops in which senior managers present a view of the future, and workshop delegates are invited to comment, discuss, and look at the implications for their own areas of work, as well as feedback upon the ideas given to them. This cascade process needs to involve the whole management structure and may consume large quantities of resource.

(iii) The Sprint - time scales are of vital importance in implementing AMT. Change usually comes about in

Our methodology (Fig.ii) begins with an audit of the current company business situation and manufacturing strategy, which incorporates a clear assessment of what is going well, what opportunities there are, and where problem areas exist. Sometimes this audit is the response to a crisis, sometimes the response to new knowledge identified with a "champion" of a particular cause. Inevitably, this audit involves the exploration of various alternatives along a variety of competitive dimensions (cost, productivity, quality, lead time, etc) and forms the basis for the development of manufacturing strategy.

Developing a manufacturing strategy frequently demands that senior manufacturing managers fundamentally rethink their approach to manufacture. It normally involves reappraising what to buy and what to make, capital investment in equipment, parts rationalisation, relationships with suppliers, factory layouts, design to manufacture, inventory policy, production management methodologies, etc.

In this way, the methodology begins with ideas and concepts about business and manufacturing strategy in relation to environmental opportunities, pressures and constraints.

Once top management has agreed the current situation and a way forward, the methodology moves into a "design for delivery" phase in both organisational and technological terms. Organisationally, the methodology involves both restructuring and culture change. The aim is to bring about an organisational delivery system designed to exploit the new technology to the full and contribute to the agreed end-state. At the same time, production managers, engineers and computer scientists need to be designing a technological horizon to deliver the manufacturing strategy.

Finally, the methodology is delivered into the company by the development of three interrelated support strategies. These three are outlined in Figure (ii) and involve on the one side a strategy for technological installation, on the other side, a strategy for structural and cultural change within the company. Changes in the latter are brought about by changing reward and control systems on the one hand, and by devising a training and development strategy on the other. Nowhere is the importance of the latter better illustrated that in the implementation of MRP II, where attitude change as well as understanding is vital if the new system is to avoid problems of data capture

<u>(iii)</u> <u>The Sprint</u> - time scales are of vital importance in implementing AMT. Change usually comes about in short bursts of upheaval followed by periods of relative stability. Pacing the change can be as important in success as the content of the change itself. Sprints may be said to have the following characteristics:

(a) They are managed from the very top of the company.

(b) They are focused on one or several key dimensions of competitiveness central to the survival and development of the company, e.g. cost reduction, quality improvement, lead time reduction, etc.

(c) They involve all members of the company and are seen to do so.

(d) They have measurable targets of success, e.g. costs remain the same for the next x months whilst AMT is installed.

(e) They have a fixed time frame in which targets must be achieved.

The sprint is the focus for company-wide activity in managing strategic change inside the organisation. Sprinting is an activity which can engender much support and create much enthusiasm within the company. However, sprints consume much resource,but may not be sufficient if change is to be sustained over the medium and long term and the organisation is not to slip back into crisis once the initial burst of energy created by the sprint. The aim of the sprint is to bring an ailing company back "up to speed" and make it competitive once more. It is the prime mechanism for bringing about step change.

<u>(iv)</u> <u>The Performance Ratchet</u> - If sprinting encapsulates the idea of step change, then the performance ratchet encapsulates the idea of continuous improvement.

The energy, commitment and enthusiasm demanded from management and organisation members is difficult to sustain over a protracted period. Indeed, the rate of change during the sprint would itself prove a major stumbling block to organisational effectiveness were it to continue unabated for a any length of time. Once the company is launched on its new path, with the accompanying technological advance, a further mechanism is required to ensure that the technology is exploited to maximum business effect. The performance ratchet does this by ensuring that minor incremental continuous changes are made by those most appropriate to initiate and support them.

The performance ratchet usually takes the form of senior management requesting from operating divisions or sections of the company the kind of performance improvements that they (the operating divisions or sections) believe that they could initiate and develop over the next time period, coupled with estimates of the degree of certainty that these performance improvements might be achieved. These may or may not be approved depending on how well they fit into an overall business plan, but the establishment of a performance ratchet such as this has the effect of continuing the impetus for change, and involving those concerned with the delivery of change in defining the targets. Performance ratchets are not the operating plan of manufacturing departments. They are concerned with change rather than delivery, and provide the link between strategic direction and operations during times of continuous improvement.

These four activities form the practical basis of an implementation methodology for managers.

7. REFERENCES

1. Berger, A.W., 1977,'Successful M.R.P. Installation the first time', 20th Annual APICS Conference; Proceedings, Cleveland, Ohio, 194.

2. Bessant, J., and Haywood, W., 1985, 'The introduction of FMS as an Example of CIM', Brighton Polytechnic.

3. Bodenstab, C.J., 1970, 'Ten Tips for Successful Implementation of Computer Systems', Financial Executive, 38, 64-70.

4. Ingersoll Engineers, 1985, 'Hi-tech can Flop in Factories', Guardian (12.5.85).

5. Meredith, J.R., 1981, 'The Implementation of Computer based Systems',Journal of Operations Management, 2, 11-21.

6. Meredith, J.R., and Hill, M.M., 1987, 'Justifying New Manufacturing Systems: A Managerial Approach' Sloan Management Review, Summer, 49-61.

7. Smith, J.S., and Tranfield, D.R., 1987, 'Strategic Methodology for Implementing Technical Change in Manufacturing'- University of Warwich, Conferenceof the British Academy of Management.

8. Smith, J.S., and Tranfield, D.R. 1988, "A Catalytic Implementation Methodology for CIM, International Journal of CIM Vol.1., No.4, forthcoming.

9. Tranfield, D.R., and Smith, J.S., 1986, 'Managing Technological Change: Tackling Taken for Granted Assumptions' - 3rd International Conference on Human Factors in Manufacturing - IFS Publications.

10. Tranfield, D.R., and Smith, J.S., 1987, 'Training for smooth revolutions: a methodology for technical change'. International conference on Training for Change - the Revolution in Commerce and Industry: YEE - Proceedings, 283, London.

11. Tranfield, D.R., and Smith, J.S., 1988, 'Managing Rapid Change' Management Decision Journal Vol.20, Spring.

12. Voss, C., 1985, 'Success and Failure in AMT" Working Paper, University of Warwick.

13, Waterlow, J.G., and Monniot, J.P., 1986, A Study of the State-of-the-Art in Computer-Aided Production Management - ACME Directorate of SER

Chapter 4

Buyer/supplier relationships: a tool for analysis and improvement

D. K. Macbeth, L. F. Baxter, N. Ferguson and G. C. Neil

1. INTRODUCTION

This paper describes work being done at the Glasgow Business School into Managing Buyers in an AMT Environment, Macbeth et al (1). AMT (Advanced Manufacturing Technology) is another of those acronyms subject to misunderstanding. Indeed the original ACARD report refers to "any substantial, relevant and new manufacturing technique the adoption of which is likely to lead to changes within a firm in manufacturing practice, management systems and approaches to design and production engineering of the product". [ACARD 1983(2)]

It is in this context that the new technology or philosophy of Just-in-Time Manufacture became the focus of our work. Schonberger (3)(4)

1.1 JIT Theory

Just-in-Time is predicated upon four main concepts, i.e., perfect quality, making the material flow, waste elimination and people involvement. It does this by providing an umbrella over a number of well known but (in the West) not previously stressed approaches. These draw on industrial engineering and method study approaches to reduce set up times, thereby enabling small batch production economically. Good quality avoids the need for protective buffers of stocked material and preventive maintenance and unstressed processes avoid interruptions in the flow through machine breakdowns. Much of this is enabled by involving people in the decision making and improvement activities at the work place.

1.2 Research Process

Our team has worked with five major Buyers, some sixteen of their suppliers, to develop case studies of existing practice in Buyer/Supplier Relationships. This information has been further refined in a larger cross-section of the Electronics and Mechanical Engineering Sectors by means of extended interview and questionnaire data. This data has provided us with the means of classifying existing relationships, as well as indicating ways in which organisational activities might be modified to some benefit.

2. WHY THE SUPPLY CHAIN?

In examining the Buyer/Supplier Relationship we have chosen to react to one of the facets of JIT which suggests that closer relationships with fewer suppliers is a goal that Buyer Companies should be setting themselves. In Shapiro's words "the firms .... that were most effective were so because of their ability to gain competitive advantage based on establishing important relationships with suppliers .... relationships which enabled their companies, in a fundamental way, to compete effectively". Shapiro (5)

2.1 Internal/External JIT, Sequential/Parallel Development

There is a belief in some quarters that there is a difference between what goes on inside the parent company and what goes on in the Supply pipeline leading into that factory. Perhaps this is simply a reaction to these managers looking for a quick fix and pushing their inventory problem back onto the suppliers. Certainly, there is little point in getting deliveries four times a day if it takes three weeks for the goods to pass through the incoming inspection and the incoming goods store. In fact, the argument goes further and develops a pattern of JIT implementation in which JIT purchasing is the last thing to be tackled. This suggests a sequential dependency where none exists. In our view JIT impacts so many areas of a manufacturing business and requires modifications in so many directions that unless attention is given, in parallel, to improving the Supply Chain, then the good things achieved internally will not be fully realised.

2.2 Manufacturing Deliverables

One of the authors has argued elsewhere that manufacturing has to achieve on Quality, Delivery and Cost - the Manufacturing Deliverables. [Macbeth (6)] The terms include within themselves a number of particular considerations, the essence of the argument is that Quality underpins and enables everything in manufacture (or Business generally?).

Given this foundation then, what acts as the metronome is Delivery, but to every customer along the chain until the ultimate paying customer is satisfied. Cost therefore becomes an outcome of correct decisions and performance on the other two and should not be managed as an entity in its own right.

Of course, inherent in the waste elimination goal of JIT is the final destination of these three, i.e., Perfect Quality, Instantaneous Delivery and Minimal Cost, and certainly to make JIT Supply work, quality levels have to be at historically incredible levels, delivery reliability is paramount and speed of response becomes more important,while cost must always be a concern, but defined in a new way - total acquisition cost (Neil et al 7) and targeted to reduce in real terms.

3. AN IMPROVEMENT MODEL

This is best demonstrated by means of a diagram - see figure 1.

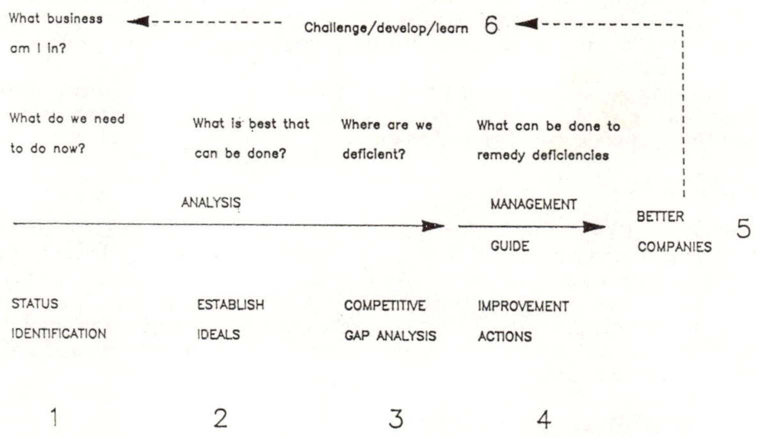

Fig. 1 Improvement model

This model recognises that each business has certain unique features by virtue of its position within a supply chain in a certain industry in a certain location. It also recognises that the strategic dimension within the company influences much of what happens.

The company also has a set of minimum specifications which it understands are necessary to survive and prosper in its current market place. This will largely be specified in its contracts with buyers, but it also covers the need to understand what the company needs to do internally to satisfy its own objectives in meeting the customer's expectation. These two constitute phase 1 of a 6 phase model, where the intention is to identify existing status. Phase 2 recognises that a "best practice" performance exists somewhere. The establishment of these "ideals" can come from three directions. The JIT ideals referred to earlier could form a target but in some ways are unhelpful, except as an indicator of direction, not as a measure of currently achievable performance. The second source can be from the kind of research activity we are currently engaged in, but this is of an aggregated, more general, nature. What is required is an industry, product and component specific set of data on current best practice, probably best performed as a rolling programme of benchmarking, not simply on technical capability, but in total manufacturing operations.

Phase 3 allows the comparison of phases 1 and 2 to identify existing deficiencies and therefore recognise those gaps in performance in which competitors can gain an edge. It is recognised, however, that the source of the problem and/or the possible solutions may not lie exclusively in either the supplier or the buyer domains. As a result, phase 4 provides indicators of possible company specific or joint initiatives to reduce the size of the competitive gaps identified. It must also be recognised that alternative routes to a particular destination may be possible and attractive.

Success in this process provides the means for more competitive companies, but phase 6 is the beginning of the virtuous circle of improvement.

4. POSITIONING TOOL

The activities identified as phases 1 and 2 above are focused by means of a tool of analysis and classification developed at Glasgow and called by us the Positioning Tool. This has as its purpose the location, within the multidimensional space, of a relationship and the production of a series of metrics on a graphic aid to highlight the phase 3 situation.

4.1 Features Incorporated

The Positioning Tool develops from 3 broad Dimensions, each composed of 3 Factors, which can have a variety of topics impinging on them generated by a large number of questions grouped into 9 main sections.

Dimensions and Factors form the basis of the graphic aid, as indicated in Fig. 2 later.

There the concept is that the 3 Dimensions of **Capability, Requirements** and **Performance** are interrelated one with another and externally up and down the Supply Chain.

Capability is the company's raw ability to meet demands placed on it.

Requirements are the demands placed on the company by its customers or by the company on its suppliers and/or internally on itself.

Performance relates to how well the company has converted its raw Capability into actions to meet the Requirements placed upon it.

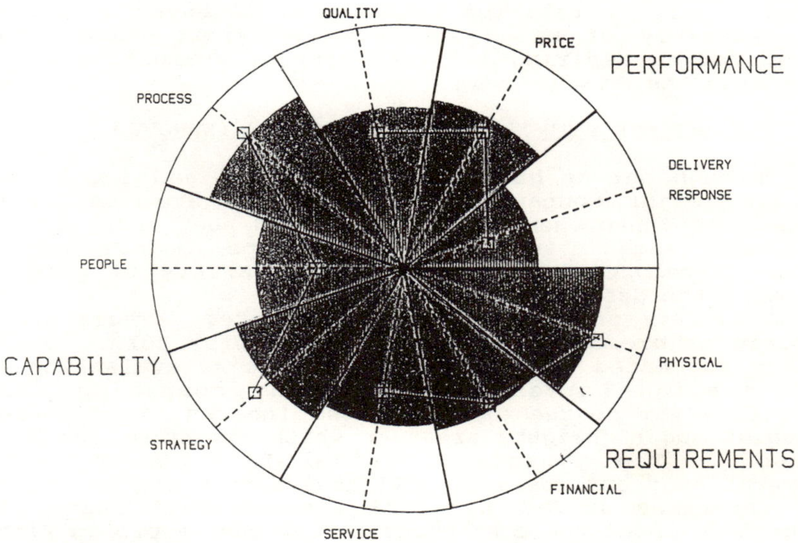

Fig. 2 Positioning Tool

The graphic shows the subdivision of the 3 Dimensions into each of the relevant Factors. By means of scoring answers to individual questions which contribute to topics which influence Factors we can create a pattern of a given relationship at a point in time compared with some ideal. Thus it becomes easy to identify the relative gaps in competitive ability. Note, however, that the scoring is not an absolute measure but is relative to the ideal and serves as an indicator of problem areas that managements should address. Thus in the diagram Delivery response may be less than it should be, but for a variety of reasons. It could be, for example, that schedule stability or forward demand visibility from the customer is inadequate or that excessive change-over times are the norm in the supplier, or that the second level supplier cannot yet react at the same rate as the main supplier. In each case the root cause can be traced back through the information structure and possible action points and managerial choices highlighted.

4.2 Method of Use

At present the Positioning Tool is created following the answers given to a wide ranging questionnaire administered by our research team. These are analysed by us off site before feedback to the companies. When finally developed the tool should be largely capable of self-administration in a variety of modes. These will include an existing Buyer/Supplier relationship where both parties seek improvement and guides to action priorities; a supplier seeking to evaluate its position within a market sector, i.e., in relation to a number of buyer companies; and, possibly of major interest in import substitution terms, for an aspiring supplier company evaluating how to break into the Supply Chain.

5. IMPROVEMENT PATHS - ROUTE MAPS TO BE IMPROVED

Through our on going case generation and involvement with our major company collaborators we are identifying examples of improvement paths which have been successful in addressing some of the competitive gaps. This will be further developed to demonstrate the variety of possible routes to the desired relationship state.

A variety of routes is to be expected. There are no panaceas or prescriptive solutions on offer here. Rather what is suggested is a process of analysis, evaluation and communication between participating companies which recognises the degree of uniqueness inherent in different companies and highlights areas on which attention should be focused. The management guide which is the major deliverable of our research will assist this process in the ways described above, but will also make it easier for managers to embed the continuous improvement approach within their normal activities, for the tool can be used as a reference point in evaluating progress towards the ideals. Since current best practice is also developing the analysis and improvement action planning and implementation will also be on going.

The need to embed the process is already evident from our work where our evidence would suggest a wide recognition of certain aspects of JIT activity, but a failure of many supplier companies to convert that awareness into understanding and action within their own organisations and especially within their own suppliers. Our other findings to date are detailed in Baxter et al 1988 (8), Ferguson et al 1988 (9), Macbeth et al 1988 (10).

6. FURTHER DEVELOPMENT

This work was conceived as a paper based exercise, but elements of the application of the data gathering instruments, the generation of the Positioning Tool scores and the graphic aid, are all areas where software support is indicated, while aspects of the management guide might also lend themselves to the advisor mode of knowledge based systems. These possibilities will be examined, but our immediate priority must be to ensure that the paper solutions fit the needs of our industrial collaborators in the first instance and are seen to be relevant more generally.

The concept of benchmarking to establish current best practice is one that is worthy of co-ordinated effort probably on a series of industrial sector bases and would provide both a standard for comparison and, by attention to performance on each of the measures outlined above, provide companies with an on going stimulus for improvement.

7. CONCLUSION

Just in Time manufacturing is not a passing fad. Elements of it may not be immediately appropriate in particular organisations, particularly if it is regarded as a quick fix solution. The underlying principles apply in any organisation and can show dramatic improvement in a relatively shorter time span and with less obvious investment, i.e., the investment is in people, training and organisational change, rather than in high technology capital "solutions".

The difficulty lies in companies actually beginning the move to better manufacturing, since the typical company is full of "opportunities" to improve.

The Positioning Tool provides an aid in prioritising the improvement activities and by its nature supports and enables moves towards a more integrated approach to Supply Chain management. We look forward to the opportunity to develop it further with interested parties.

References

1. Macbeth, D.K., et al, 'Managing Suppliers in an AMT Environment', Grant Number GR/E/12337 ACME Directorate of SERC, Swindon, England.

2. ACARD, 'New Opportunities in Manufacture', HMSO, 1983.

3. Schonberger, R.J., Japanese Manufacturing Techniques: Nine Hidden Lessons in Simplicity, Free Press, New York, 1982.

4. Schonberger, R.J., World Class Manufacturing: the Lessons of Simplicity Applied, Free Press, New York, 1986.

5. Shapiro, R.D., 'Towards Effective Supplier Management: International Comparisons', Working paper 9-785-062, Harvard Business School, Boston.

6. Macbeth, D.K., Strategy and Management of Advanced Manufacture, Technical Communications, Letchworth, England, 1989.

7. Neil, G.C., Baxter, L.F., Ferguson, N., and Macbeth, D.K., 'The Continuing Importance of Cost in Supplier Selection in Support of JIT', Proceedings 3rd International Conference on JIT, IFS (Publications) Ltd., Kensington, England, 1988.

8. Baxter, L.F., Ferguson, N., Macbeth, D.K., and Neil, G.C., 'Materials and Information Flow: Managing the Supply Chain', Factory 2000 International Conference, I.E.R.E., Cambridge, 1988, 81 - 86.

9. Ferguson, N., Baxter, L.F., Neil, G.C., and Macbeth, D.K., "Better Management of the Supply Chain in Support of JIT Manufacturing Operations", IIE International Systems Conference, St. Louis, 1988.

10. Macbeth, D.K., Baxter, L.F., Ferguson, N., and Neil, G.C., 'Buyer - Vendor Relationships with Just-in-Time: Lessons from US Multinationals', IE, September 1988, 34 - 41.

Chapter 5

Intelligent rule based simulation of flexible assembly systems

A. K. Kochhar and P. Rayner

ABSTRACT

This paper describes a rule based system which has been developed to aid the manufacturing engineer in the adaptation of existing flexible assembly systems for new or improved products. The rule based system can be used to carry out a simulation to check whether a new product can be assembled on an existing system and what changes, if any, must be made to the configuration of the assembly system. The actual simulation is carried out in two stages involving a rough check and a detailed check. These checks are concerned with the possible conflict between the assembly requirements of a product and the functions performed by an existing flexible assembly system. A final check compares the simulated cycle time with the required cycle time.

1. INTRODUCTION

It is difficult, at first, to understand why companies should be contemplating the automated assembly of batch type products when one considers the advantages of manual assembly. Miller (1) has pointed out the advantages of manual assembly: "Automated assembly has a formidable competitor in manual assembly which requires little capital investment and whose flexibility and dexterity are hard to emulate in a machine." Such statements can discourage the attempts to develop a flexible assembly system. It is only when the disadvantages of manual assembly are considered that automation comes into its own. Manual operators very often suffer from fatigue, sickness, and assemble products of non-uniform quality. There is also the problem of labour relations. In addition, some very intricate assembly operations cannot easily be performed by manual operators, particularly in the electronics industry. In some instances there are obvious economic advantages of automation over manual assembly. For instance, dedicated assembly systems perform assembly operations much quicker than humans and there are distinct economic advantages as a result. Dedicated assembly, as the name suggests involves the assembly of one particular product. Many companies are unable to take full advantage of dedicated assembly mainly because they do not produce the volumes of single product types (i.e. batch type assembly). Usually the equipment utilisation needs to be high in order for that equipment to be economically justified. Hence these companies must introduce assembly systems which can handle a range of products, with quick changeover times between runs. These type of assembly systems are known as general purpose assembly systems or more commonly flexible assembly systems.

2. DESIGN OF FLEXIBLE ASSEMBLY SYSTEMS

As would be expected, no assembly system has yet been designed to assemble all conceivable products. This is because assembly problems are too different. Although the flexible assembly systems developed so far can sometimes deal with many products, it must be realised that in the majority of cases these products belong to the same product family and are very similar.

Flexible Assembly systems can be divided into two different categories, (i) Assembly Cells and (ii) Assembly Lines. The first category involves products with relatively few operations (typically less than 15 operations). Products which require more operations tend to be placed into the assembly line category, mainly due to cycle time requirements. Assembly lines can be regarded as a number of individual assembly cells linked together by some form of transfer mechanism (e.g. conveyor, AGV etc.). Many attempts have been made to generalise assembly system design (2,3,4,5). Most of the work so far has concentrated on new assembly systems (i.e. "green field site" applications). Little work has been performed on the adaptation of existing assembly systems for new or improved products.

It is necessary to carry out two major tasks when assembling any product, namely, parts presentation and assembly.

Parts presentation can be split further into parts feeding and parts orientation. In flexible assembly systems it is necessary to ensure that the parts presentation is performed automatically. There are many different types of parts presentation devices in existence today based on a whole host of mechanical actions (e.g. vibratory, rotary, reciprocating etc.). Some components are suitably packaged in a pre-orientated form and the assembly system can take advantage of this (e.g. pallets, magazines, bandoliers etc.). All of these feeders are dedicated feeders one normally associates with dedicated automated assembly, i.e. parts presenters which feed and orientate one component only. In general purpose assembly dedicated parts presenters are still used when the component being fed is present in all of the product variations and is added at the same point in the assembly. In recent years 3 other types have been developed: (i) multi-part feeders, (ii) sensor based feeders and (iii) programmable feeders. These 3 categories of feeder are ideally suited for general purpose assembly.

2.1 Part Feeding

The multi-part feeder (6) has been developed at the University of Salford. The concept of this feeder is to change the orientating devices everytime a different component requires feeding. The feeder is a linear vibratory feeder and has any combination of 2 to 6 part dependent tracks. The tooling is machined from solid bar material and is easy to change, the estimated track changeover time being approximately one minute. The feeder can handle small parts only with maximum dimensions of 30 x 30 x 30 mm.

Various sensor based feeders have been developed in the past and many of these are based around artificial vision systems. One such system developed at Hitachi (7) combined a multi-level bowl and a vision system. The system has been used in conjunction with a robot in the assembly of a portable tape recorder consisting of 13 parts. Swift and Dewhurst (8) have suggested a neon helium laser with an electro-optic detection system for the recognition of parts. Parts are fed along a bowl feeder track and pass through a viewing area. Scattered light intensities are generated and compared with a component signature. The advantage of this type of feeder is that it can recognise slight asymmetrical features which are too small to be identified by conventional bowl feeder tooling.

The best known programmable feeder has been developed by Westinghouse Research and Development Centre (9). This feeder consists of 2 belts, an inclined feeding belt and a horizontal return belt. Parts are orientated by means of a group of custom made orientating devices mounted on an indexable turret. Different components can be fed by indexing to the required orientating devices.

2.2 Assembly

Assembly operations can be categorised into:

- tasks requiring insertion and
- tasks not requiring insertion (manipulative tasks)

Insertion tasks can be difficult to perform and many parameters can affect the success or failure of robotic insertion, e.g. the presence of part chamfers, the type of fit etc. Insertion can be simplified by designing compliance into the assembly system, whether it be in the robot arm/wrist or in the worktable (10). Since there are a number of different robot configurations, the choice of robot to perform a number of assembly tasks is crucial in the design of assembly systems. The robot must:

- be capable of reaching all of the components presented to it
- possess the dexterity to perform all of the assembly tasks
- satisfy the load carrying requirements

- be compatible with other equipment in the system (i.e. controller interfacing)
- satisfy the speed requirements
- be capable of providing the necessary insertion forces AND
- be accurate enough to position components

The robot is the most flexible part of any general purpose automated assembly system. However this flexibility can be greatly reduced by the choice of end effector. End effectors can be tools (e.g. screwdrivers, soldering irons etc.) or grippers. In general, grippers tend to be dedicated to one or a few components. Hence turret grippers or gripper changing may be needed to maintain the system flexibility. Both of these approaches tend to increase the assembly cycle time, thus resulting in the need for a trade-off between flexibility and speed.

Typically a flexible assembly system would consist of one or more work stations at which individual opeations are carried out to assemble the required product. The design, development and implementation of flexible assembly systems is an expensive and time-consuming process. Due to the relatively high cost of implementing such systems, it is necessary to maximise their utilisation.

To achieve this objective, it may be necessary, in some cases, to use the Flexible Assembly System for assembling new products for which it was not designed in the first place. Before any assembly can take place, it is necessary to assess the suitability of the flexible assembly system to carry out the necessary tasks (11).

3. RULE BASED SYSTEM APPROACH

Rule based systems are increasingly being used to help make decisions in a variety of environments. Expert systems have been developed to help design new flexible assembly systems for a particular product family (12, 13). No attempt has been made to create tools for examining the potential use of existing assembly systems for assembling products other than those for which they were designed in the first place.

When faced with the problem of deciding whether or not an existing flexible assembly system would be capable of assembling a new product, which may be completely new or a variation on the existing products, a number of issues need to be considered.

- General Assembly System Configuration (i.e. Cell or Line)
- Equipment Issues (i.e. parts presenters, end effectors and robots)
- Product Related Issues
- Spatial Requirements

The first three of the above issues can be investigated using heuristics, (i.e. IF/THEN rules) and reasoning, and are difficult to solve using conventional programming methods. These issues can also be subject to uncertain and unreliable information. For example, consider a robot which is required to insert a component into another component involving a very tight clearance fit. The uncertain knowledge about this task may involve the robot's ability to position the components prior to insertion. In order to increase the probability of successful insertion it would be advantageous to machine chamfers (or to increase the size of the chamfers) on one or more components. A rule based system approach can account for uncertainties of this kind by the use of its architecture and the use of logical relationships. FASES (Flexible Assembly System dependant Expert System) is a rule based system which attempts to emulate human reasoning when investigating the suitability of an existing assembly system to assemble a new or improved product.

The final issue of spatial requirements has not been included in the system since there are very complex simulation models which account for spatial requirements. Simulation packages such as GRASP have been developed which address spatial problems within robotic assembly cells.

4. STRUCTURE OF THE RULE BASED SYSTEM

The technical aspects of the flexible assembly system simulation are covered by 2 spearate checks, a ROUGH CHECK and a DETAILED CHECK. The ROUGH CHECK is a filtering process and its purpose is to dismiss the obvious unsuitability of the assembly system for assembling a given product and to inform the user that changes need to be made. The filtering process is carried out by a number of parametric checks (e.g. dimensions, weight etc.) on component requirements and equipment capabilities.

The ROUGH CHECK examines the general suitability of the layout to match the number of work stations required with the number of work stations included in the selected system.

If the general layout of the system is found to be suitable, the simulated system cycle time is compared with the required cycle time. This is followed by an investigation of the equipment at each station in the flexible assembly system. The suitability of feeding devices is checked by examining whether a part has been fed by any of the feeders in the past. The potential use of robots is examined by considering the combined weight of a component and its gripper. The shape and size of components are also checked against the capabilities of a potential feeder or gripper etc. Following the completion of the rough check, the user is informed of the results. Each piece of information is displayed on a separate page. Relatively few pages of results infer a reasonable match whereas many pages of results indicate a poor match. Should a totally incompatible product/assembly system be under investigation, the gross incompatability will soon become apparent by the large number of changes suggested at the end of the ROUGH CHECK. The ROUGH CHECK takes the form of a conventional program with dedicated procedures and is written in the artificial intelligence language POP11.

DETAILED CHECK

The filtering process of the ROUGH CHECK gives the user an insight into the potential of a suitable match being found. The DETAILED CHECK, as the name suggests, is a more rigorous investigation into the suitability of a product/flexible assembly system combination. The DETAILED CHECK deals with tasks that are/can be performed by programmable equipment. The rule based Detailed Check system consists of a knowledge base, a detailed data base and an inference engine.

The Knowledge base contains all the rules relating to the DETAILED CHECK and is represented in the form of a decision tree. The database contains detailed information relating to the individual elements which are used at the work stations forming part of a flexible assembly system, and the product to be assembled. During a simulation run the database is expanded with the information relating to the product and the assembly system. A decision tree represents the IF/THEN rules by assigning a node within the decision tree to a rule consequence. Linked to this node are other nodes which form the factors (or other rule consequences) which make up the rule. The node can be an AND node or an OR node, for example

```
IF      Factor B     AND Factor C
             are satisfied                   AND node
THEN    Goal A is satisfied

IF      Factor Y     OR   Factor Z
             is satisfied                    OR node
THEN    Goal X is satisfied
```

The inference engine has been designed around a task independent knowledge based system called TREE, developed at the University of Bradford (14). Procedures from TREE have been used to form FASES, the task dependant rule based system. The inference engine contains many facilities such as:

- uncertain responses
- assessment of the consequences of uncertain responses
- interrogative responses such as "WHY" and "HOW"
- consequences of the user giving an unfavourable response
- explanation facilities

The purpose of these facilities is to provide the USER who may be a non-expert with a tool to examine a potentially complex problem.

Man/Machine Interface has been designed to provide the USER with a "friendly" environment in which to work. The rule based system is menu driven and the USER is prompted for data or answers to questions.

FASES has 2 modes of operation:

(i) the input of rules/data
(ii) performing a simulation

In the first mode of operation i.e. data entry, the USER is allowed to describe a flexible assembly system, or part of it, and also a product or component. The USER is given the option of describing new data or modifying existing data. When the details of at least one assembly system and one product have been input a simulation can be performed. In order to perform a simulation the filenames of the datafiles containing information about the product and assembly system need to be input along with various production data (e.g. production volume required, number of shifts to be worked, number of hours per shift etc.).

A modular approach is used for describing a Flexible Assembly System. The assembly systems are considered as being made up of a number of "building blocks". As shown in Figure 1, assembly systems are made up of a number of assembly workstations and similarly workstations are made up from pieces of equipment (e.g. feeders, robots etc).

Since it may be possible that other assembly systems are built from similar pieces of equipment, this modular approach means that there is no repetition of data. There is no need to re-enter the equipment data since it is already present.

Similarly, products can be considered in a modular fashion, i.e. product families are made from products, and products are assembled from individual components.

5. EXAMPLES OF RULES USED

As already described the DETAILED CHECK takes the form of a decision tree and is concerned with the investigation of the suitability of flexible equipment for a particular product. The most flexible of all assembly equipment is the assembly robot which may be used, for example, to insert a male component into a female component with a clearance fit. Part of the decision tree applicable to this problem is shown in figure 2. The rules for this problem, which form part of a much larger rule base, are as follows.

Rule (1)

```
IF      Likelihood of positioning for the        (Rule 2)
          clearance fit.
   and  Likelihood of suitable dexterity         (Rule X)
THEN    Suitable ROBOTIC INSERTION FOR
          CLEARANCE FIT.
```

<u>Rule (2)</u>

```
IF      Certain positioning for the              (Rule 3)
          clearance fit
  or    Conditional positioning for the          (Rule 5)
          clearance fit
  or    Probable positioning for the             (Rule 6)
          clearance fit
THEN     Likelihood of positioning for the
          clearance fit
```

<u>Rule (3)</u>

```
IF      Maximum Error of the positioning         (Factor a)
          system is less than the minimum
          clearance between the 2 parts
  or    [the parts are suitably chamfered        (Rule 4)
  AND   there is compliance within the           (Factor b)
          assembly system]
THEN     Certain positioning for the clearance
          fit
```

<u>Rule (4)</u>

```
IF      one or both of the components is/are     (Factor c)
          chamfered
  and   maximum error of the positioning         (Factor d)
          system is less than the minimum
          combined component chamfer clearance
THEN     the parts are suitably chamfered
```

<u>Rule (5)</u>

```
        Certain positioning IF

IF      the clearance between the 2 components   (Rule V)
          is suitably increased
  or    Certain positioning if compliance        (Rule W)
          is built into the system
  or    Certain positioning if chamfers          (Rule X)
          are machined
  or    Certain positioning if compliance        (Rule Y)
          built in and chamfers machined
THEN     Conditional positioning for the
          clearance fit
```

<u>Rule 6</u>

```
IF      Probability of successful insertion
          for a clearance fit = Pc
  and   the probability Pc is acceptable to
          USER
THEN     Probable positioning for clearance fit
```

6. USER INTERACTION

As mentioned earlier, the knowledge based system has 2 modes of operation. The first mode allows the USER to input data and/or rules, whereas the second mode of operation allows the USER to perform a simulation. In the first mode of operation the USER can DEFINE, MODIFY, DELETE or LIST datafiles which contain information about products, components, assembly systems or parts of the assembly systems. In the second mode of operation the USER is prompted for the names of 2 files, one containing data about a particular product and one containing data about a flexible assembly system. The USER, at this point, is asked to confirm the correctness of data in these datafiles prior to the working database being constructed.

The information relating to the product and assembly system is transferred into the working database and the USER is prompted for further information such as annual volume required, number of shifts/day to be worked, number of hours per shift, number of days/year available and efficiency expected of the assembly system. From this information the system calculates the required assembly cycle time. The USER is given the opportunity to assign tasks to stations, alternatively the system will assign tasks based on line balancing techniques. The USER is informed of the general layout suitability (i.e. number of stations required vs. number of stations available) and is given the option of changing the required cycle time to fit the simulated cycle time, should there be a discrepancy.

The USER now has two options (i) continue with the ROUGH CHECK or (ii) change the data already input. If the USER decides to continue with the ROUGH CHECK the internal workings of the program are not shown to the USER until the 'Results' page(s) are displayed on the screen. Depending on the number of changes which are recommended by the ROUGH CHECK the USER may wish to continue with the DETAILED CHECK. During detailed check, the USER is required to interact with the system on a much more regular basis than for the ROUGH CHECK. This interaction takes the form of question and answer session with the USER often being asked to respond in a pre-determined fashion, (i.e. the USER may be asked to type 'y' for a favourable response to a question or 'dK' for an uncertain response to a question etc.). The USER may also be asked to type a numerical value. The system also supplies the USER with the option of questioning decisions or in fact questioning why a particular question is being asked. The USER can also ask for explanations at this stage of the simulation. The decisions reached during the DETAILED CHECK depend heavily on the USER'S reaction to the questions asked. On completion of the DETAILED check, the USER is supplied with recommendations regarding the suitability of the chosen assembly system for the selected product.

7. CONCLUSIONS

This paper has described the structure and operation of the rule based system, FASES, which can be used to assess the potential suitability of a Flexible Assembly System for assembling a given product. Such flexible assembly systems will become increasingly important when manufacturing companies have to switch quickly between the assembly of different products.

The implementation of automated and flexible assembly systems involves a very considerable expense. When a flexible assembly system has been implemented, it must be used within the company for a time period greater than its payback period. In many cases, it would mean that an existing flexible assembly system should be used to assemble a far wider variety of products. The rule based system, FASES, provides the manufacturing engineer the information required to assess the suitability of an existing assembly system for assembling different products along with advice on any changes which must be made. In many cases, positive assessment could result in the prolonging of the working life of the assembly system.

Ths knowledge based system should have a wide applicaton as technology moves forward to develop a whole host of 'soft' equipment to replace the traditional hard equipment, and also as assembly systems become more flexible (adaptable) in their design.

8. ACKNOWLEDGEMENTS

The authors are grateful to the Science and Engineering Research Council for the award of a Research Studentship to Paul Rayner. The technical support and financial assistance of Lucas Engineering and Systems is gratefully acknowledged.

REFERENCES

1. Miller, J., 1984, 'Robotic Assembly System Design, Cost and Manufacturing Viability', I. Mech. E. Conference Proceedings, 'UK Robotic Research', London, England.

2. Boothroyd, G., and Dewhurst, P., 1983, 'Design for Assembly Handbook', Department of Mechanical Engineering, University of Massachussetts.

3. Scott, P., and Husband, T., 1984, 'The Search for Cost Effective Robotic Assembly', I. Mech. E. Conference Proceedings, 'UK Robotic Research', London, England.

4. Kahler, S., Andreasen, M.M., and Ahm, T., 1983, 'Design of Assembly Systems for Maximum Flexibility'. Proc. of the 4th Int. Conf. 'Assembly Automation', Tokyo, Japan.

5. Lotter, B., 1985, Assembly Automation, 5, 2, 103-109.

6. Redford, A., and Lo, E.K., 1986, 'Robots in Assembly', Open University Press, Milton Keynes.

7. Suzuki, T., and Kohno, M., 1984, 'The Flexible Parts Feeder Which Helps Robots to Assemble Automatically', In: Hartley, J., Flexible Automation in Japan', I.F.S. Publications Ltd.

8. Swift, K.G., and Dewhurst, R.J., 1983, 'Versatile Automatic Handling Systems based on a Laser Electro-optic Component Recognition Technique', Proc of 15th C.I.R.P. Int. Seminar 'Manufacturing Systems', Amherst, Massachussetts, U.S.A.

9. Cowart, N.A., 1980, 'Programmable Assembly Research, Technology Transfer to Industry'. Phase 2, Year End Report, NSF Grant ISP78-18773, Westinghouse R & D Centre.

10. Arai, T., and Kinoshita, N., 1981, Assembly Automation, 1,4, 201-210.

11. Ma, X., and Kochhar, A.K., 1988, Proc. I. Mech. E., 202, B1, 51-59.

12. Simsik, D., Kovac, J., and Madrasz, L., 1988 ' An application of expert systems in the design of robotic assembly lines', Proc. I.E.R.E Conference 'Factory 2000', Cambridge, U.K.

13. De Winter, D., and Van Brussel, H., 1985 'An expert system for flexible assembly system design', Proc. Automan 85 - 3rd European Automated Manufacturing Exhibition and Conference', Birmingham, U.K.

14. Kochhar, A.K., Burns, A., Pegler, H.C., and Pain, A., 1988, 'Knowledge Based Design of Manufacturing Systems', Proceedings of the SERC - ACME Research Conference, Nottingham, U.K.

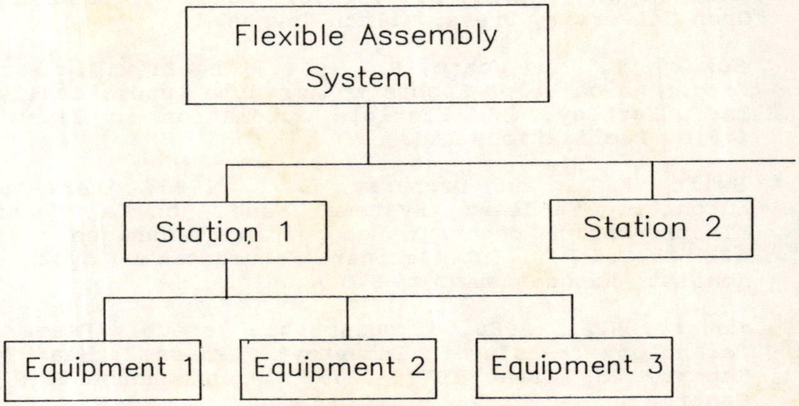

Figure 1. Modular Representation of a Flexible Assembly System

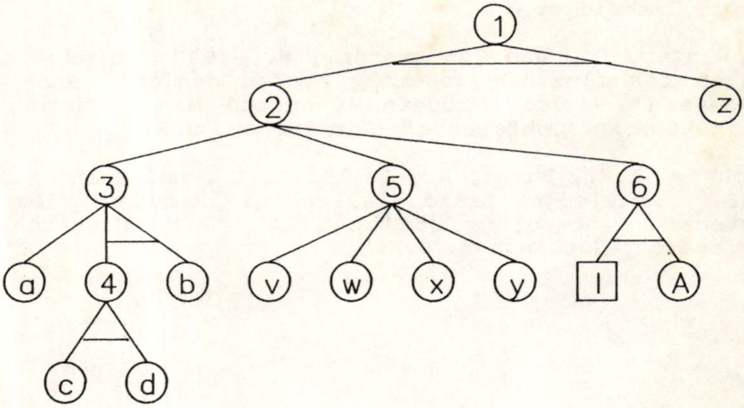

Figure 2. Schematic Example of the Rule Structure Used in the Detailed Check

Chapter 6

Analysis techniques for manufacturing applications —the production planning function

G. M. Williams and P. J. Sackett

SUMMARY

The authors describe an application of structured techniques to the analysis of the Production Planning function in an aerospace company. This work was undertaken in two distinct phases. The first phase modelled the functions of the scheduling system, examining its structure and operations at a broad business level. The second phase involved a detailed analysis of the data flow and data structure in the Master Production Scheduling function. An assessment of the merits of appropriate techniques for manufacturing applications is presented. The models were treated as an accepted description of current operations, and as such were used as the focus for discussion about the development of Production Planning as a whole, and Master Production Scheduling in particular. The issues highlighted during the development and review of the models are discussed, as are aspects of the particular models. The discussion focuses on the definition and development of the Master Planning function, exploring issues associated with the integration of Manufacturing and Commercial functions.

1.0 INTRODUCTION

The work described forms part of an SERC sponsored CASE project, undertaken by The CIM Institute, Cranfield and Plessey Aerospace Ltd., Titchfield [SERC 85/501276 - Computer Integrated Components Management Systems]. The practical benefits obtained by the Company in sponsoring this work on Production Management Systems are highlighted. The paper describes an application of structured techniques to the analysis of the production planning function in an aerospace company. It clarifies systems analysis in this context and identifies the methods and support tools available. It provides a means of assessing these techniques for manufacturing applications. A description of the use of structured hierarchies and conceptual models as part of the same project is provided elsewhere (Sackett and Williams (1)).

2.0 THE PROBLEM

The complexity of the practical production scheduling problem is great, and for some companies quite literally overwhelming. The organisation and management of material flow into and through a factory provides a basis of cost control. The means by which this management and control is performed is all important; efficient and effective management providing the basis for a competetive edge. The dynamic nature of manufacturing industry, particularly in terms of competition and technology poses special problems. Technological panaceas for an organisational problem do not exist (2).

The intellectual challenge posed by the practical scheduling problem has attracted researchers , from a wide range of disciplines. Their work has been directed particularly towards job shop and, more recently, Flexible Manufacturing System scheduling. However, much of this work has been of limited benefit to the practitioner. Browne (3) states, and Ballacher (4) reiterates, *'the scheduling problem, as it is posed in the literature, does not exist in real job shops'*. This proposition is reinforced by the authors own experience. Talking about production management, Burbidge (5) has said:

'it has to be admitted that Production Management is at present a backward science. It has failed first, to develop a universal framework which covers Production as a whole, and shows how its parts are related, and second, it has failed to develop any body of universally accepted laws to guide future decision making. I believe systems theory provides the best hope for a solution of these problems.'

Companies have difficulty identifying when it is appropriate to undertake a particular course of action or adopt a particular philosophy of operation. This difficulty is compounded by problems associated with clearly defining both the state of current operations, and what changes may be required. Prescriptive solutions are too frequently applied in companies which can readily identify the symptoms of a problem without an understanding of their root causes (Puttick(6)). The dangerous nature of such an approach is well documented (Krupp(7) and Kamenetzky(8)). One example in the authors direct experience is provided by a US Airframe Manufacturer, which spent $1m on an MRP system, a further $46m and 2 years customising it to their own particular requirements. The company is still dissatified with the resulting system which has contributed little to its competitive performance.

Recent research work has been aimed at defining operating structures and characteristics, in order to provide a frame of reference for researchers and practitioners alike. The definition of a control hierarchy can provide a company with a sense of structure, and a perspective on its operations; the hierarchy reflecting the functions performed and the decisions taken, not the mechanisms which are used to exert or communicate that control. Such a definition can help to reduce the scale and complexity of the manufacturing 'problem'. However a more detailed description of the company is required to provide a basis for improving or integrating operations, and for comparison with existing generic models. Systems analysis is a means of developing such descriptions or models.

3.0 SYSTEMS ANALYSIS

Systems Analysis is widely understood to apply to the design and analysis of software systems; the traditional Data Processing role. This definition does not cover the full scope of systems analysis. The analysis of any system can be undertaken for reasons other than the development of a specific piece of software (Checkland(9)). It is useful to examine what is meant by Systems Analysis, by defining a system and its life cycle. From there it is possible to discuss the differing viewpoints that can be taken during the analysis of a system and identify their practical applications.

3.1. Object and Information Systems

Manufacturing systems can be regarded as ill structured or soft systems. This research work has drawn a clear distinction between the types of system that found in a manufacturing organisation (Van Koetsveld (10)). Without such an understanding it is difficult to select the appropriate tools to support the analysis of a system, or to fully appreciate the relationship between the systems being analysed.

Object System : Part of the real world, example; a manufacturing plant or company.

Information System : Derived from an object system and supporting it through functions that may be automated, computerised or manual, example; a manufacturing information system.

Software System : Supports specific processes of an information system, example; a Materials Requirements Planning system.

Hardware System : Supports the operation of specific tasks within an information or software system, example; a computer or robot.

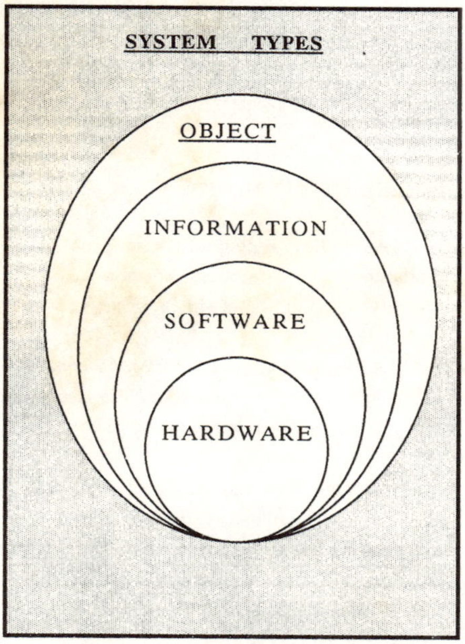

3.2. System Development Life Cycle

Understanding the System Development Life Cycle ensures that the analysis technique chosen will address the required stages of systems analysis, design and development. It is now a recognised means of viewing the development of any software system, but can equally well be applied to other system types. It has eight stages of development;

0. Preliminary Investigation
1. Problem Definition
2. Logical Design

3. Physical Design
4. Construction
5. Integration & Testing
6. Installation
7. Evaluation, Maintenance & Support

The role of analysis is generally biased towards the earlier stages of the development cycle. This dictates that the quality of the analysis effort has a fundamental influence on the overall development of the system. A lack of structure in the analysis and design stages of the System Development Life Cycle has been identified as leading to major problems in overall system development (Ross (11) and Couger et al(12)).

Analysis problems highlighted include ;

* System/Requirement Mismatch
* Lack of Maintainability & Reliability
* Delivery Overruns
* Cost Overruns

3.3 Structured Analysis

An understanding of these problems , and their causes, has led to the development of methodologies to address specific parts of the System Development Life Cycle, and support specific views of a system.

'A methodology is a recommended collection of philosophies, phases, procedures, rules, diagramming techniques, documentation and controls which will allow a structured and clearly defined route from a problem statement to a solution.' (Yeomans(13))

Starting from an origin in software program design a wide variety of techniques (Bravco and Yadev(14), Post(15), and Cameron(16)) have since been developed to cover different problem requirements. The role of such techniques has grown from merely defining software solutions (programs) to fully analysing problem areas. It can be seen that a range of general requirements for a good methodology can be identified when compared with the systems problems highlighted earlier (Fitch(17)).

In general terms the following judgement criteria should be considered;

* Scope of subject matter covered by the methodology
* Ease of use
* Ease of understanding of its output
* Adequacy for analysis subject matter
* Effectiveness of methodology, ie does it support good systems development.
* Does it support an integrated approach

4.0 METHODS AND SUPPORT TOOLS

Three different but related viewpoints can be taken of any system. These being ;

* Functional Analysis - Activities & Interrelations
* Information Analysis - Data & Relations
* Dynamics Analysis - Behaviour over time

4.1. Functional Analysis

A functional analysis involves the definition of the activities and interrelationships within a system. The objective of such analyses is to provide coherent, formal and structured representations of complex systems, avoiding the pitfalls of ambiguity and inconsistency. The functions of a system can be defined initially at a high or generalised level, and then subsequently decomposed into a series of related sub-functions. Such an analysis is thereby capable of providing the desired level of detail without the need to represent the full complexity of the system in a single description (Bravco and Yadev(18)).

IDEF-0

The first of a three method set in the ICAM Definition Methodology, 'IDEF-0 is a method used to produce function models of a system' (19) . It was developed as a subset of Structured Analysis and Design Technique (SADT) - a method developed by D.T.Ross of SofTech Inc..

The concepts behind the IDEF-0 functional modelling methodology are :

1. The provision of clear and concise documentation.
2. Good communication through :
   - Simple diagrammatic representation
   - Texts and Glossaries in natural language
   - Gradual exposition of detail through hierarchical decomposition
   - Node charts
   - Limitation of detail (& complexity) at each level
3. Maintain rigour and precision
   - Detail exposition control at each level
   - Bounded context
   - ICOM code auditing
   - Syntax, semantics and diagramming rules
   - Purpose, context and viewpoint
4. A defined methodology : step by step procedures for providing modelling, review and integration tasks.
5. Model function not organisation. Functions are relatively stable compared to organisations which are company dependent.

4.2. Information Analysis

Information analysis has its roots in software system design, where a wide variety of techniques have been developed to improve the productivity and quality of software production. Starting from the most basic flow diagram definitions in the 1960's more formalised approaches to information analysis have been developed during recent years. Two distinct approaches to information analysis can be found in current methodologies, distinguished by their end objectives. One approach uses data analysis as the initial stage of a methodology to provide a software solution, and is represented by Structured Systems Analysis and Design Methodology (SSADM). This takes a series of structured steps from initial analysis through to detailed physical design and coding. A second approach uses data analysis as an integral part of an object system analysis, along with functional and dynamic analyses. It is represented by IDEF-1, which is the Information Modelling Methodology in the IDEF method set.

The analysis techniques of SSADM were used in this project because :

- Documentation and computer support tools were readily available.
- SSADM is a CCTA (UK government) standard, and is used in several Plessey businesses.
- SSADM provides greater potential for short to medium term improvements in DP productivity.

SSADM

SSADM is an amalgamation of the basic work and principles of Bachmann, Codd, Gane & Sarson(20), Jackson(21), and Yourden and Constantine(22). It is a comprehensive methodology tracing a path from initial study through to program specification. Its structure comprises six basic stages. Each stage is broken down further into a number of steps, each step employing specific techniques, standard documentation and producing defined outputs. The major analysis techniques used in SSADM are :

Data Flow Diagrams (DFD) : Define an operational picture of a system, describing how data moves around the various processes and data stores of the system, and between the system and external entities. The technique provides a simple top down diagrammatic structure allowing a detailed analysis to be undertaken without losing sight of the overall system operation.

Entity Model : Describe the logical data structure of a system in terms of its entities (data groups) and their relationships (access path). In defining the existence of the data and its generic relationships there is a validation of the data flows and data stores described in the DFD's.

Entity Life History (ELH) : Provides a view of data by showing the behaviour of individual entities against time. They can be used to describe both normal and exceptional life cycles. Each entity on an Entity Model will require an ELH. Each DFD can be validated to ensure it supports the events of the ELH.

4.3. Dynamics Analysis

Dynamics analysis describes the time-varying behaviour of manufacturing systems in such a way that the descriptions can be analysed using computer simulations to generate measures of system performance (Bravco and Yadev(23)). The work undertaken on dynamic analysis is covered in detail elsewhere (Williams(24)).

5.0 THE PLESSEY AEROSPACE APPLICATION

The objective was to identify means of maximising the benefits of cellular manufacturing and MRPII for the sponsoring company. The broad nature of the overall objective meant that the project would encompass all functions impinging on production scheduling and control, as well as the core functions. The approach adopted followed the now well established approach of defining and reviewing logical abstractions of current (AS-IS) and desired (TO-BE) operations, and the means of getting from one to the other in practical terms (Ward(25), and Wang and Smith(26)). Rather than follow the whole method through this paper concentrates on the analysis techniques required to support this approach.

5.1 The Production Planning Function and IDEF-0

A generic Plessey Company model was used to help establish the context of Plessey Aerospace production scheduling analysis. The purpose of the model is to provide a complete and concise description of current scheduling activity, an AS-IS model, with a view to identifying any scope for improvement. Twelve man weeks effort, involving interviews with 35 members of staff, produced a 42 node model decomposed through five function levels.

- A0 Run Plessey Aerospace
  - A1 Manage Business
  - A2 Design & Develop Products
  - A3 Plan & Support Production (Diagram 1a & 1b)
    - A31 Plan & Support Manufacturing Process
    - A32 Plan Production
      - A321 Create & Maintain MPS (4 Sub-Fn)
      - A322 Plan Material Requirements (5 Sub-Fn)
      - A323 Report Capacity Requirements (5 Sub-Fn)
    - A33 Sequence & Schedule Production
      - A331 Allocate Mtl & Load Factory (4 Sub-Fn)
      - A332 Schedule & Progress Prod'n (5 Sub-Fn)
      - A333 Plan Work Centre Load (4 Sub-Fn)
  - A4 Manufacture Products
  - A5 Deliver & Support Products

The model was accepted as a valid description of company's 1987 scheduling activity. It identified several minor inconsistencies and duplications in operations, which were speedily rectified. More importantly it highlighted 2 major opportunities for improving the effectiveness of the planning functions ;

1. A requirement for an improved mechanism to support finite capacity shop floor scheduling, proving a link between MRPII and Cell Communications systems.

   The Genetik Planning Board package is being used to develop an appropriate shop scheduling support tool, using the IDEF-0 as the basis of the Functional Specification.

2. A requirement to improve the Master Planning function in terms of its defined role, support mechanisms and channels of communication with the business. Implicit in this was the need to balance the trade off between MRP philosphy and company budgetry control.

   The complex and pivotal role of the Master Scheduler was seen to require further investigation. The IDEF-0 model was used to set the system boundaries for this Data Analysis.

5.2 Master Planning and SSADM

The initial IDEF-0 study identified the Master Planning function as a significant informal element in the overall manufacturing information system. It suffered from the absence of adequate software system support, despite the operation of an MRPII system. The role of the Master Scheduler within this function was poorly defined, with a cross departmental split of authority and responsibility.

An 8 man week data analysis was performed, using Data Flow Diagrams as the principal modelling techniques, to define processes, data flows and data stores. These were cross referenced with Entity Models, Entity Life Histories and the company Data Dictionary, using the Design Master Analysis support tool from ICL. The result was a highly detailed description of current practice, which was consistent with existing software system descriptions.

The diagramming and cross referencing was made easy by the use of the analysis support tool, so the major effort was expended on capturing the informal elements of the system. The temptation to stray away from a strict AS-IS description into a TO-BE design was significantly greater than with the earlier functional analysis. However, the accuracy of the final model proved useful in the later design of both the formal manual mechanisms and the required software support systems.

6.0 CONCLUSIONS

The application of systems analysis techniques to manufacturing applications provides a rigorous and structured examination of production planning operations. The results of the modelling process have provided a vehicle for considerable discussion. They provided an agreed platform from which to start developing the production planning function, and a basis from which to review conceptual models as a part of that development.

The project has provided a valuable insight into current Production Planning activities, defining them in terms which provided a clear progression to the implementation of improvements.

REFERENCES

1. Sackett, P.J., and Williams, G.M., 1988, ' An Integrated Structure for Production Scheduling and Control', Proc. of IERE Int. Conf. ' Factory 2000 - Integrating Information and Material Flow', Cambridge, England.

2. Anon, 1988, 'The Culture Shock of Automation', Financial Times, 7/10/88.

3. Browne, J., Boon, J.E., and Davies B.J., 1981, ' Job Shop Control', Int. J. Production Research, Vol.19, No.3, pp.633-643.

4. Ballakur, A., and Steudel, H.J., 1984, ' Integration of Job Shop Control Systems : A State of the Art Review', Journal of Manufacturing Systems, Vol.3, No.1, pp.71-79.

5. Burbidge, J.L., 1985, ' Production Planning and Control : A Personal Philosophy', Computers in Industry, No.6, 477-487.

6. Puttick, J., 1987, ' Fads and FashionsThe Engineer, 8/10, 64-65.

7. Krupp, J.A.G., 1987, ' MRP Re-Implementation',Production and Inventory Management, Fourth Quarter, 73-81.

8. Kamenetzky, R.D., 1985, ' Successful MRPII Implementation can be Complemented by Smart Scheduling and Sequencing Systems' Industrial Engineer, October, 44-52.

9. Checkland, P., 1981, ' Systems Thinking, Systems Practice', John Wiley & Sons, ISBN 0-471-27911-0.

10. Van Koetsveld, M.J., 1984, ' A Study of the Usefulness of IDEF-0', CAM-I Report B-84-184, Poole, UK.

11. Ross, D.T., 1977, ' Structured Analysis (SA) : A Language for Communicating Ideas', IEEE Transactions on Software Engineering, Vol.SE-3 No.1.

12. Couger, J.D., Colter, M.A. and Knapp R.W., 1982, ' Advanced Systems Development/Feasability Techniques', Pub: John Wiley & Sons 1982, ISBN 0-471-03141-0.

13. Yeomans, P., 1984, ' Modelling Systems Architectures Using the IDEF Methodologies', CAM-I productivity Seminar - 22/10/84.

14. Bravco, R.R. and Yadev S.B., 1985, ' A Methodology To Model The Information Structure of an Organisation', The Journal of Systems and Software Development, No.5, 59-71.

15. Post, J., 1986, ' Application of a Structured Methodology to Real-Time Industrial Software Development', Software Engineering Journal, November, 222-235.

16. Cameron, J.R., 1986, ' An Overview of JSD', IEEE Transactions on Software Engineering, Vol.SE-12 No.2, 222-240.

17. Fitch, N.G., 1987, ' An Analysis of Integrated Manufacturing Systems', MSc Thesis, Cranfield Institute of Technology, UK.

18. Bravco, R.R. and Yadev, S.B., 1985, ' A Methodology to Model The Functional Structure of an Organisation', Computers In Industry, No.6, 345-361.

19. Anon, ' ' ICAM Architecture Part II- Volume IV - Functional Modelling Manual (IDEF-0)", Report AFWAL-TR-81-4023, Pub: Softech Inc.

20. Gane, C. and Sarson, T., 'Structured Systems Analysis : Tools and Techniques', ISBN 0-13-880328-5.

21. Jackson M.A., 1983, ' System Development', Pub: Prentice Hall International, ISBN : 0-13-880328-5.

22. Yourden, E. and Constantine, L.L., 1979, ' Structured Design: Fundementals of a Discipline of Computer Program and Systems Design, Pub: Prentice Hall Inc., ISBN : 0-13-854471-9.

23. Bravco, R.R. and Yadev, S.B., 1985, ' A Methodology to Model the Dynamic Structure of an Organisation', Informations Systems, Vol.10 No.3, 299-317.

24. Williams G.M., 1988, ' Simulation as a Tool for the Analysis of Manufacturing Systems', Working Paper : AMT005/88.

25. Ward, R.J., 1985, ' Guidelines to a Methodology for Planning Computer Integrated Manufacturing', CAM-I Report R-85-EAP-02.

26. Wang, M. and Smith, G.W., 1988, ' Modelling CIM Systems Part 1 : Methodologies', Computer-Integrated Manufacturing Systems, Vol.1 No.1, 13-17.

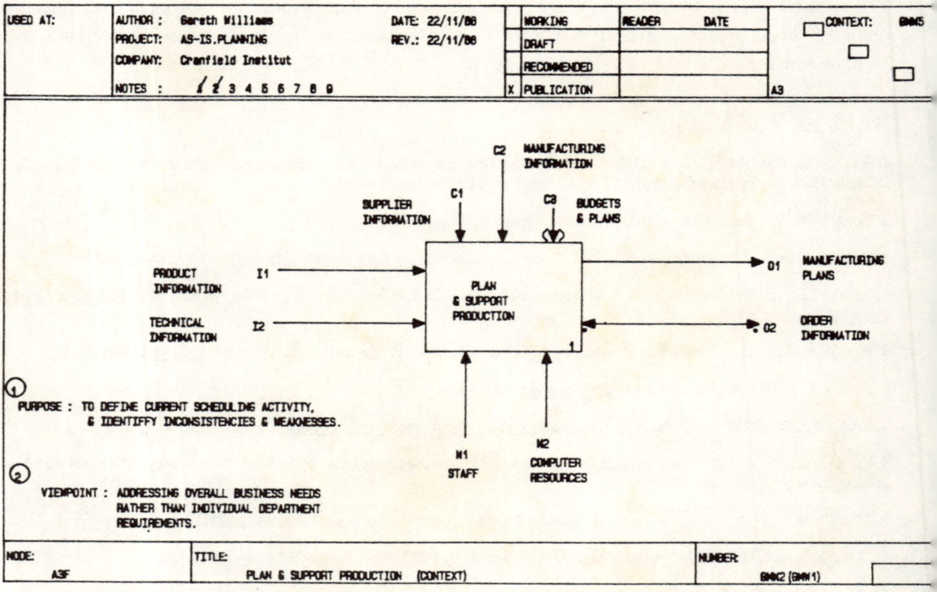

DIAGRAM 1a: Plan & Support Production - IDEF-0

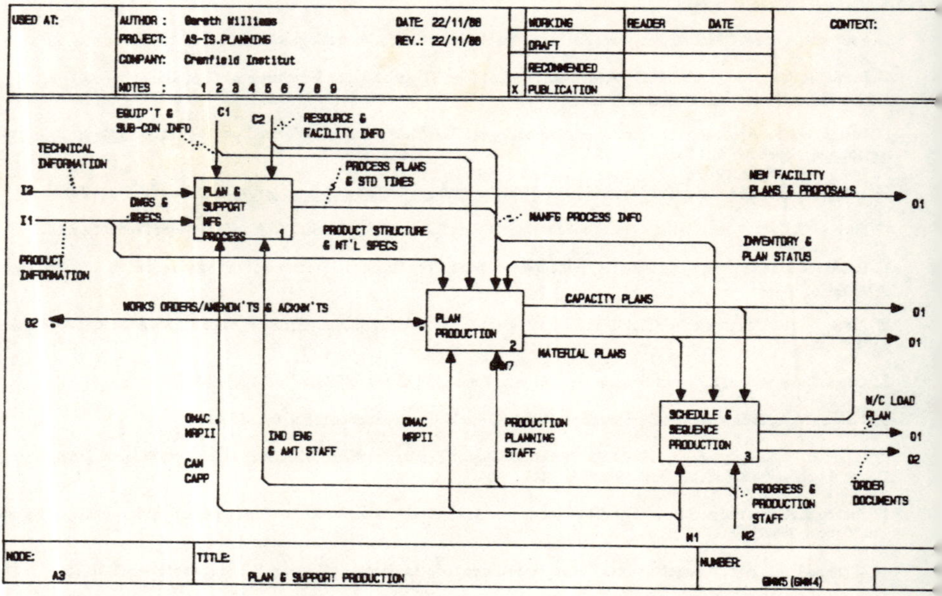

DIAGRAM 1b : Plan & Support Production (Decomposed)

Chapter 7

Buildings for advanced manufacturing technology

Dr. J. B. Harris

Synopsis

This paper describes a two-year investigation into the performance of industrial buildings in relation to the changing functional requirements of advanced manufacturing technology (AMT). Studies of 24 AMT installations in the UK were compared with similar facilities in Sweden, West Germany, Italy and Japan to compare the problems encountered elsewhere and to assess the overall levels of architectural quality and performance achieved across a wide range of batch-production facilities in differing industrial buildings.

The findings of the investigation, together with the conclusions and recommendations, have been published in the form of a Research Report and a Briefing and Design Guide, supplemented by a seminar programme to encourage their implementation.

Background

The European economy depends to a large extent on manufacturing engineering. This sector of industry produces and assembles a vast range of components involving millions of workers in thousands of companies. Since 1977, onwards, the manufacturing environment of the medium-volume, medium-variety batch production sector - usually about 70 per cent of the total - has been transformed by advanced technology. Traditional 'machine shop' equipment such as individual milling, boring and turning machines has begun to be replaced by groups of computer-controlled, multi-function machines and 'flexible manufacturing systems'. The newest of these operate on the work-piece, change their own tools and transfer the machined parts from point to point using robotic transfer lines and automatic guided vehicles, all under computer control and economical for small batch production.

Developments of this kind have significant effects on industrial buildings, but these had not been systematically studied, not had the effects of installing AMT in conventional factories been investigated previously.

Objectives

The objectives of the investigation were as follows:-

To investigate the problems which have arisen in introducing advanced manufacturing technology into industrial buildings in the UK in comparison with similar facilities in Scandinavia, West Germany, Italy and Japan.

To determine to what extent conventional industrial structures can satisfactorily accommodate AMT and to what extent new structures and facilities are likely to be required.

To explore the feasibility of a Briefing and Design Guide for buildings for manufacturing engineering.

Methodology

The investigation began with the carrying out of a general review of the development of AMT using conventional literature search procedures. This review served to define the field and to identify the significant companies, manufacturers and institutions to be consulted. From this information a selective list was prepared of the important installations to be studied in the UK and abroad. A draft questionnaire was also prepared, to form the basis for structured interviews and to order the resulting data. The questionnaire was tested in a pilot study of six representative companies and some amendments made to it. Finally, the main study was carried out, in three phases: companies accessible from the university in one day were visited first; secondly, longer-distance companies involving over-night accommodation were grouped into several 'round-trip' tours and, finally, the foreign visits were grouped into four installations in each of four countries.

Findings

The installations ranged in size from one or two machining centres with simple transfer systems to sophisticated installations with complex AGV routings and extra facilities.

The buildings also varied widely, in age, size and structural type. But the significant differences were between buildings in the UK and those elsewhere.

In the UK almost all the installations were in existing, ageing buildings, some dating back to the 1930s. These outdated structures had produced a series of problems for works managers and AMT installers: space for access and for the installation was often restricted; the construction of the buildings was often sub-standard, demanding expensive up-grading; there were problems with environmental and machine services, and with maintenance and safety. Companies had introduced AMT into their buildings only at the price of some difficulty in terms of time, management stress and additional expense.

In Europe, the buildings were newer and larger, with wider column spacings and greater heights. Consequently, companies had been able to introduce AMT with minimal spatial restrictions and fewer practical difficulties than companies in the UK. There was also a high level of quality in the shop-floor working environment showing that a serious concern for the highly-skilled work-force was an important element of company policy. In some cases, too, the building and the manufacturing process had been designed as an integrated man/machine/building system, the preferred installation layout, its servicing patterns and communication links forming the basis for the new building design.

In Japan, the buildings were also newer and larger; some factories had been designed on two floors to economise on site area - an idea which might prove useful on congested sites in the UK. Moreover, Japanese companies used their buildings to promote their image and to influence potential customers by means of viewing galleries, taped commentaries and relaxation facilities.

Conclusions

From these findings it is clear that AMT installations impose new requirements not only on existing factories - their location so far in the UK - but also on the new factories of the future. In the UK particularly, the physical attributes of industrial buildings had significant effects on the case of installation and subsequent operation of the manufacturing facility.

The phrase 'conventional industrial structures' is a misleading generalisation about a building type which is by no means homogeneous. Conventional industrial structures in the UK constitute a series of differing architectural forms which vary widely in dimensions and types of structures.

In these circumstances, the extent to which Britain's existing factories can satisfactorily accommodate new AMT installations can only be understood in fairly general terms. But it is clear that conventional industrial structures in the UK are only suitable for the long-term introduction of advanced manufacturing technology to a limited extent, and not without considerable inconvenience, expense and delay. Moreover, in comparison with their foreign competitors, a significant proportion of the UK's manufacturing industry is being seriously hampered by outdated industrial buildings on congested sites.

There is a need, therefore, for new approaches, new design concepts, and a new attitude to industrial buildings, for several reasons: firstly, industry is re-organising more frequently, and re-allocation of space occurs more often. The required adaptability will need to be 'designed-in' in a systematic way. Secondly, the key, educated work-force of modern, knowledge-based industry will prefer companies which provide a good working environment. Thirdly, many manufacturing processes are becoming 'self-contained' and therefore cleaner. New buildings can enhance this improvement better than elderly structures. Fourth, in a competitive world, companies need to present an efficient corporate image to give customers confidence and maintain market status. Finally, foreign competitors are already importing higher design standards by commissioning their own premises in this country.

Recommendations

The primary recommendation is that AMT installations should not be designed in isolation but as one element in an overall process/people/ building system. Moreover, the interactions between the operators, the installation and the building should be as carefully considered as the individual elements of the manufacturing facility.

Where an existing building is to be used, a systematic appraisal of its nature and condition should be made. This should include spatial and environmental performance in comparison with likely future standards structural type, capacity and condition, and the extent of available services for the building and the installation.

The relationship between the manufacturing process and the industrial building must be borne in mind from the outset in terms of:

1. Spatial and dimensional requirements.
2. The construction of the building envelope.
3. The environmental and installation services.
4. Safety and security.
5. Maintenance of the building and plant.

Designers should ask a series of questions to ensure that the factors influencing the design, construction and operation of the facility are taken into account from the beginning. (A set of typical questions is included in the published report).

Implementation of the recommendations

The award of the grant attached particular importance to the dissemination of the findings of the study, so as to encourage the adoption of the recommendations by industrialists and designers.

This has been achieved in three ways: firstly by the publication of a Research Report which describes in detail the problems which may be expected to occur in the installation of AMT, and sets out the findings, conclusions and recommendations of the study.

Secondly, a Briefing and Design Guide has been prepared. This explains the importance of including the industrial building as a prime factor in all AMT proposals; it shows how each phase of implementing an AMT proposal has architectural implications; it provides a briefing check list of questions to be answered, specific design recommendations and criteria, and plans of noteworthy layouts.

Thirdly, the first of a series of seminars to explain the research and its recommendations has been held at the University of Strathclyde. The event was well received by participants from British Aerospace, Rolls-Royce, GEC, Plessey and Hewlett Packard, amongst others. Further seminars are planned for the future. The Research Unit continues to offer a design and consultancy service on all aspects of architecture for manufacturing industry.

The full report of the project: Advanced Manufacturing Technology and the Industrial Building: a Research Report with a Briefing and Design Guide is available, free, from The Architecture for Industry Research Unit, School of Architecture, University of Manchester, Oxford Road, Manchester M13 9PL or by telephoning 061 275 6936.

Recommendations: some examples.

Installations should have adequate height available so as not to restict preferred layout. An optimum clear height of 6.0m should be aimed for.

If the installation is to form part of the publicity programme of the company, consideration should be given to viewing platforms and overhead walkways for visitors. This factor may also influence the location of the facility within the factory.

No permanent heating in the high-bay stores, but a temperature of 10°C to be maintained to prevent condensation and weekend freezing.

Lighting levels in the high-bay stores to be 250 lux at the control stations and 20 lux in the gangways.

In new buildings the structural grid of columns should be not less than 20m x 20m, and preferably larger in at least one dimenson. In existing buildings, closer spacings of columns are likely to lead to spatial and servicing difficulties and the removal of some columns may be worth-while.

Installations should be planned to provide good internal access for delivery and maintenance using overhead cranage or a mobile crane of 15 tonnes capacity.

Lighting level for the overall monitoring of the process from the control room to the 200 lux. Lighting level for repair and maintenance 'down-time' periods to be 500 to 750 lux.

Temperature-sensitive facilities should be located away from large glazed areas, or shielded from them. To keep within the tolerance of production cells and the inspection abilities of coordinate measuring machines a temperature of 20-22 degrees centigrade should be maintained; variations of +/− 5 degrees should be avoided.

Factory floor heating rates should be monitored and adjusted to compensate for the heat output of the AMT installation. An adequate supply of cool, clean air must be available around the machine air intakes; local heat extractors or air cooling units may be required in relatively enclosed spaces.

Factory-floor ventilation rates should be re-assessed when a new AMT installation comes into operation so as to compensate for any build-up of air-borne contaminants such as misted oil-based coolant or fine metal dust.

Adequate height should be available for ancillary facilities such as automated vertical tool stores, and operators' control rooms. Any 'high bay' storage should be considered separately.

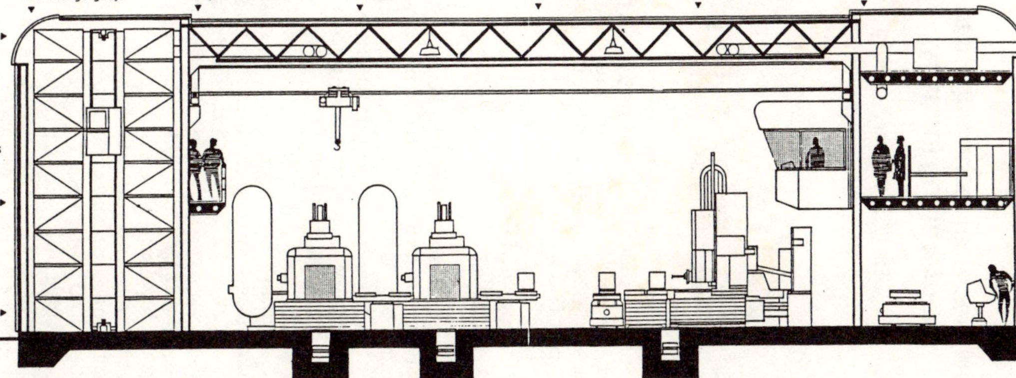

Typical floor slabs should have a compressive strength of 30-40 Nmm² with a thickness of 300-400mm; this may need to be increased to 500mm, or individual foundations provided, for larger machining centres.

Ground floor slab reinforcement must be no closer to AGV guidance wires than 600mm on either side and at least 75mm below the finished floor level.

The planar tolerance of the finished floor surface should be no greater than 2mm over 100mm, 4mm over 1.0m, and 10mm over 4.0m for rail-guided systems. For AGV systems, maximum gradients should be no greater than 7 degrees with a planar tolerance of +/− 5mm over any 2.0m distance.

A minimum aisle width of 3.5 or an optimum width of 5.0m should be provided in close proximity to the machine cells.

The requirements of a centralised swarf-processing facility should receive particular attention in respect of the plant itself and its associated ductwork which should avoid AGV routes. Where this is not possible, removable bridges over the duct, with integral wire guides and snap connectors may be provided.

Chapter 8

How new electrical technology can profit manufacturing industry

Dr. D. J. Walker

1. Introduction

In the industrial countries of the Western World the historic link between primary energy consumption and economic growth has been decisively broken. Between 1974 and 1985, in spite of a small drop in overall energy consumption, the Gross Domestic Product (GDP) of the OECD countries increased by 27%. In contrast a close correlation still exists between GDP and electricity useage (up 30% during the same period). For the United Kingdom the corresponding figures are an 18% rise in GDP between 1974 and 1985 which was accomplished using 5% less total primary energy but 12% more electricity.

Electricity is also important to the future well being of the economy in that 58% of the energy we currently use is in the form of oil and gas, both shortlived and finite fuels. Increasingly electricity generated from coal or nuclear energy will form a growing proportion of the energy resources of the nation.

The strong links between the use of electricity and economic performance are further enhanced by the work of the Electricity Council Research Centre at Capenhurst near Chester. Capenhurst is the Research Laboratory of the Area Electricity Boards and it was established in 1965 to carry out research into new and improved methods of electricity utilisation and distribution. With a staff of 350 which includes approximately 100 professionally qualified scientists and engineers it is uniquely engaged in searching for electrical solutions to the problems of UK Industry. At Capenhurst the emphasis is placed on a multidisciplinary approach to problem solving with research staff from a number of technical disciplines being used on a given project as required.

Approximately one third of the research and development at Capenhurst is in the area of electricity distribution with the aims of enhancing the efficiency and reliability of the Electricity Board's distribution networks. New methods of finding faults when they occur and novel techniques for detecting incipient failure have been devised. This work has involved a diverse range of projects from establishing a lightning flash location system, assessing the corrosion resistance of metal conductors, measuring the strength of wooden poles and studies of arc furnace flicker and distribution system automation.

In an increasingly competitive world it is vital that the UK, if it is to survive as a major trading nation increases its productivity, improves its product quality and continually introduces new products which must be agressively marketed throughout the world. Capenhurst contributes to these goals in three main areas

(a) Its research work on electricity utilisation has released a flow of new technologies covering the whole spectrum of manufacturing processes from drying, metal melting and heat treatment, cutting, chemical processing, effluent treatment and product finishing

(b) Capenhurst also acts as a source of new products and processes for UK Industry to exploit under licence both at home and overseas.

(c) Finally, but by no means the least of its contributions to the economic wellbeing of the country, Capenhurst acts as a centre of expertise for the Area Electricity Boards and their customers to call on in the event of problems being encountered. In addition the facilities and expertise of Capenhurst are now available to UK Industry direct via consultancy or contract research.

The present paper will highlight some of the specific developments which will enhance the overall efficiency of UK Manufacturing Industry.

2.Drying

Throughout the UK more than 30 million tonnes of water and substantial quantities of other solvents are removed every year through a variety of industrial drying and evaporation technologies. This accounts for about 6% of the total industrial energy used. A wide variety of electrical drying techniques are already in existence and there are many instances in which a change from fossil fuels has led to energy cost savings of up to 75%. However currently new technologies are being developed at Capenhurst to extend the range of electrical options available. These include:

2.1 Rotek - The Rotary Induction Kiln

The Capenhurst Rotary Induction Kiln is an

indirectly heated rotary kiln in which the internal shell and lifters are heated by electrical induction with very high efficiency. In essence the shell of the dryer acts as the secondary winding of a transformer whose silicon iron core surrounds the shell. A centre inductor runs along the centre of the kiln and this acts as the transformer primary. Heated thin metal flights fitted longitudinally to the inner skin are also induction heated thus greatly enhancing heat transfer to wet solids. The prototype kilns developed at Capenhurst can dry either continuously or in batches, granular or powdered solids at temperatures up to 200°C and calcining tasks up to 600°C In contrast to conventional direct fired rotary kilns which require high air flows, both to support fuel combustion and to achieve good heat transfer to wet solids the Capenhurst kiln uses minimum air flow which ensures low dust losses and reduces the cost of any flue gas treatment equipment required. Successful drying tests have been carried out on products from a wide range of industrial sectors including sand aggregates, metal powders, catalyst pellets, cereal products, pharmaceuticals and clay materials.

Futher development work at Capenhurst has enabled the operating temperatures to now be extended to over 1000°C. This widens the range of solid drying and calcining tasks which can be undertaken.

2.2 **ARFA - Air Radio Frequency Assisted Dryer**

Capenhurst's work in the application of radio frequency energy to drying and other processes is well know throughout the world. The ARFA dryer uses a combination of the established techniques of air impingement or air flotation drying with radio frequency(RF) to dramatically reduce drying times. Air impingement is widely used to remove water vapour from the surface but the drying rate is often limited by the rate of mass transfer from the interior to the surface. In the ARFA dryer the application of R.F. generates heat uniformly within a material and the internal water which is particularly receptive to RF energy redistributes itself to maintain a wet surface for the hot air to remove.

The first commercial installation in the UK was for drying the leatherboard used in shoe manufacture. The ARFA dryer working with existing air impingement and RF dryers has helped to reduce drying times and led to a 50% cut in reject rate and a 15% increase in throughput. The principle of combining heat transfer mechanisms used in the ARFA dryer has now been extended by further Capenhurst research to a wide variety of operations in the food industry such as baking of biscuits and bread and the processing of breakfast cereals.

3. Heat Pumps

Over the years Capenhurst has carried out extensive research and development into the use of heat pumps for industrial applications such as drying and distillation.

For example a high temperature dehumidifier with an evaporator temperature of 80°C has been developed which has extended the range of applications for dehumidification. Among the areas where this dehumidifier has been applied are timber drying, pallet drying, and confectionery. The dehumidifier not only recovers the latent heat of the moisture evaporated from the product but also the sensible heat of the air which carries the moisture away. Dramatic energy savings have been possible, for example in one installation for timber drying the dehumidifier's running costs were 39% below those of the conventional heat and vent kiln.

Capenhurst was also involved in the heat-pump/heat recovery scheme installed at a Maltings in East Anglia. Here a horizontal cascade of heat pumps was fitted to a pair of kilns already using recuperative heat exchangers. This resulted in a halving of the drying energy costs and by replacing most of the fossil fuel plant the quantity of products of combustion passing through the bed was reduced thus enhancing product quality.

Steam compression is another area where Capenhurst expertise has found application. Industrial processes which reject large quantitites of heat at 100°C and which are steam heated at a slightly higher temperature are ideal applications for steam heat pumps. Drying, distillation and evaporation are the major processes which meet these criteria. As an example, in the distillation area Capenhurst installed the first UK closed cycle heat pump using steam as the working fluid at the Auchroisk Distillery of IDV (Fig.1). In effect the latent heat in the vapour from the still is recovered by condensing it using high temperature, 80°C, cooling water. This hot water now at 85°C is returned to a flash vessel where its pressure is reduced to 0.5 bar producing steam and water at 80°C. The water is pumped back to the condensors while the steam is compressed to about 1.8bar using an oil free twin screw compressor and then fed back to heat the stills. Total energy consumption is reported to be down by 35% and additional savings in maintenance and operating costs are reported to have arisen as a result of dramatic reductions in the demand for boiler steam.

4.Plasmatek Plasma Cutting Torch

The Capenhurst developed Plasmatek plasma torch (Fig. 2) is a major advance in industrial cutting technology. Cutters using plasma - an intensely hot inert gas - are already used for metal cutting and profiling but such cutters, in which a "transferred" electric arc is struck between the nozzle and the workpiece are limited to the cutting of conducting materials. The torch developed by Capenhurst instead produces a non-transferred arc which is struck within the watercooled nozzle to produce a narrow faster cut in any sheet material. The arc raises a jet of argon gas to temperatures in excess of 20,000°C and this emerges from the nozzle at a supersonic speed.

It has been used to simultaneously cut and heat seal many fabrics at production speeds which makes it very attractive to the textile industry and the compact lightweight design of the torch also makes it ideally suited to profiling applications. The plasma torch offers all the benefits of sophisticated laser or water jetting at a fraction of the cost. It can be used for cutting cleanly, accurately and quickly through sheet metal, glass, plastics or textiles yet both capital and running costs are much lower.

5.Ohmic Heating

Ohmic heating technology, developed at Capenhurst and licensed to APV Projects Ltd. is essentially a novel electrical method of continuously heating foodstuffs by passing a controlled electric current through the product.

Many food processes require heat for cooking and sterilisation. The use of ohmic heating generates heat directly in the product; heat transfer is thus avoided allowing difficult products to be more rapidly and uniformly processed.

Of particular interest to the food industry is the ohmic heater's ability to sterilise particulate food products on a continuous basis, prior to asceptic packaging. Most packaged foodstuffs of this type on the market today cannot be cooked or sterilised quickly because it is difficult using conventional techniques to achieve high rates of heat transfer into the solids.

Some of the more important operating benefits are perceived to be (a) It is rapid. Ohmic heating raises most suspensions of solids to the required sterilisation temperature in under 90 seconds. (b) It is gentle. There are no moving parts within the heater and heating is uniform even within the particles therefore very delicate products, for example, strawberries, can be processed. (c) There are no heated surfaces or complex shapes which would otherwise cause fouling and cleaning problems. Ohmic heating lends itself well to standard clean-in-place systems. (d) Maintenance requirements are very simple.

Amongst the foodstuffs that have been processed are prawns, carrots, potatoes, meat chunks, rice puddings soups, fruit and pie fillings.

6. The Metallurgical Industries

The metallurgical industries are amongst the largest and most intensive users of energy in the United Kingdom and Capenhurst has had extensive involvement with them.

6.1 Metal Melting

In the early years of Capenhurst a great deal of work was carried out on the production of steel using electric arc furnaces. Interest centred on the melting of prereduced iron briquettes and "flicker" problems. More recently a loop heated ladle was developed which allowed Midland Rollmakers of Crewe to expand their production at a very low capital cost. The ladle fitted with an induction loop is used for holding and transporting 20 tonnes of hot steel which is then used for the casting of large steel rolls. The use of the loop heated ladle has opened up markets previously unavailable to Midland Rollmakers.

In iron foundries Capenhurst has done much to demonstrate the advantages of coreless induction furnaces over the traditional cupola and computer programs have been developed to help foundries select the optimum size and ratings of electric melting furnaces. More recently a novel technique has been developed for the production of SG iron castings by melting selected grades of steel scrap rather than using costly low manganese imported pig iron and now the production of austempered ductile cast irons is being investigated.

In the nonferrous sector, work has concentrated on the use of coreless induction furnaces and the quality of the feedstock in determining the metal losses during melting. The method of heat transfer and the inherent stirring action in a coreless induction furnace enables metal to be melted with minimal oxidation and hence low metal losses. This is particularly important for high surface area materials such as edge trim, scalpings etc. which can give rise to high melting losses in reverbatory type furnaces because of excessive oxidation.

A high power channel induction furnace has been developed for melting aluminium. Foundries were disinclined to use these furnaces for melting aluminium in the past because of the frequent cleaning of oxide deposits from the channel walls which was required to prevent clogging. Capenhurst solved the problem by redesigning the furnace channel loop so that movement of molten metal could flush oxide away from the channel and hence prevent blockages. This approach of using electromagnetic fields to control metal flow has now been the subject of further investigation and this has resulted in an electromagnetic valve to control the metal flow during the pouring of cast iron and aluminium.

6.2 Metal Reheating and Heat Treatment

Capenhurst has considerable experience and expertise in applying induction heating to the reheating of metals for hot working and heat treatment. A high efficiency billet heater has been developed for non ferrous metals using a multi layered copper winding to reduce energy losses in the induction coil. Not only have dramatic energy savings been demonstrated but also a wider range of billet sizes can be treated in a given coil at a still acceptable efficiency.

A technique has also been developed for the continuous heat treatment of metal strip called Transverse Flux Induction Heating (TFX). In TFX instead of the induction coil being wrapped around the metal strip, it passes between two sets of opposed electrical windings which direct magnetic flux through the strip at right angles to the moving strip surface. This device has been used for the continuous heat treatment of aluminium alloy strip up to 1.2 metres wide and 0.5-1.5 mm thick. Annealing and solution treatments have been carried out but any metal can in theory be treated and TFX can also be used for continuous paint drying/curing with the added advantage of curing from the inside out rather than the outside in as happens with conventional methods of paint drying.

The use of glow discharge electron beams for metal heat treatment have also been studied in such areas as the continuous heat treatment of wire and strip, the thermochemical treatment of steels and the production of metal strip from powder.

7. Environmental Protection

Capenhurst has been active in the field of effluent

treatment and materials recycling for many years and among the wide range of products and processes developed are:-

7.1 **The Chemelec Cell**

The Chemelec Cell is designed to recover precious and toxic metals from dilute solutions such as occur in the rinse tanks on electroplating lines and in photographic processing. A fluidised bed of glass beads is used to ensure that a continuous supply of metal ions is brought to the cathode of the Chemelec Cell where they are plated out. As metal deposition progresses the plated cathodes are periodically replaced and may be reused directly as the anode in a plating tank. The Chemelec Cell has been used to recover cadmium, copper, gold, nickel, silver, tin and zinc.

7.2 **The CEER Cell**

The Capenhurst Electrolytic Etchant Recovery (CEER) Cell has been designed for the regeneration of the cupric chloride etchant commonly used in the Printed Circuit Boards industry. The production of printed circuits involves the removal by etching of a large percentage of the copper originally clad on to it.

During the etching process as the concentration of the dissolved copper in the etchant rises it becomes chemically reduced from the cupric to the cuprous state and it cannot then dissolve further copper. It must be regenerated. Conventionally this is done by adding chlorine or hydrogen peroxide which are both hazardous chemicals. In contrast the CEER Cell regenerates the etchant by oxidising it back to the cupric state and simultaneously in another part of the cell the excess high purity copper in the etchant is plated out and this can then be readily sold.

7.3 **The VO2 Venturi Aerator**

Aeration of municipal and industrial effluents is widely carried out to encourage aerobic micro-organisms to purify the effluent biologically. Both contaminated water and settled sludges can be treated in this way to reduce the levels of suspended solids and both biological oxygen demand and chemical oxygen demand. Capenhurst has developed a highly efficient venturi aerator (Fig. 3) capable of forcing 30% of the oxygen from air into true solution. The VO2 venturi aerator is a general purpose aerator, which because it offers high oxygen transfer efficiencies and good mixing of liquors should be of wide interest to industry for such diverse applications as aerobic digestion of sewage and effluents, dewatering of anaerobically digested sludges, aeration of water supplies and processes with high oxygen demands.

7.4 UV/Ozone Odour Removal

Capenhurst has discovered a highly efficient method of destroying undesirable factory smells using ultra violet light and ozone gas. Existing methods of eliminating industrial odours such as trying to mask them with a sweeter smell or by incinerating the effluent are costly and can lead to toxic by-products. Dilution is usually ineffective and absorption onto solids such as charcoal only works for some compounds and even then the resulting solids have to be either disposed of or recycled.

The device which has been developed at Capenhurst (Fig. 4) uses ozone and ultraviolet (UV) light to oxidise totally all the malodorant compounds present. The ozone dissolves in water and is converted by UV light into hydroxyl radicals. These hydroxyl radicals are reactive towards any organic compounds oxidising them to carbon dioxide and water etc. Field trials have been carried out on a range of malodorants including chlorophenols, phenols and waste derived fuels.

Discussion

Electricity is the most versatile energy source available to industry but to utilise its full potential it must be used wisely and efficiently. The electricity supply industry has encouraged this by introducing its Power for Efficiency and Productivity (PEP) award now in its fifth year. To qualify for the award scheme companies have to demonstrate that the introduction of an electrical process or service has enhanced their business and led to one or more of the following:

- A decrease in the amount of energy used per unit of production
- An improvement in product quality
- Reduced unit production costs
- Better working conditions

No less than seven of the winners of the national 1987 PEP competition featured a degree of Capenhurst involvement including four Capenhurst developed products. This emphasises the role of Capenhurst in enhancing the profitability of UK manufacturing industry by providing it with new efficient technologies for use over the whole process spectrum, by providing new products for UK industry to manufacture and market worldwide and finally acting as a source of technical expertise for the Area Electricity Boards and their customers to call on.

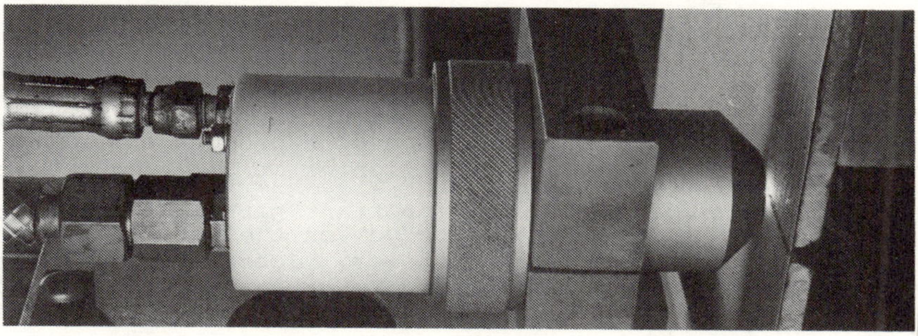

Chapter 9

The use of expert systems in process planning of turned components

B. J. Davies, A. T. Joseph and M. Kalta

Abstract

UMIST has for several years been trying to automate process planning for machined components. EXCAP S is the current state of development of an expert process planning system for turned components, which has as its input a CAD product model of the required component, and has as its output an "optimised" N-C part program.

Introduction

Process planning for machined components is the operation which converts the design of a component into detailed instructions for the manufacture of the component. These instructions include the selection of the machine, the fixturing, the cutting sequence, the cutting tools, the cutting conditions, estimates of set up and cycle times, and tool lives, production of a process plan, and in the case of N-C machines production of an N-C part program.

This operation forms an important part of the lead time for the production of a new product or the modification of a standard product to suit a customer. Jobs generally queue at process planning offices and the process, once started, usually takes from several days to several weeks . This long lead time is an increasing problem as the life cycle of products becomes shorter, caused by increasingly sophisticated customer demand and an increasing willingness and ability of suppliers to meet this demand. The life time of some consumer products such as video recorders, and cameras, is approximately equal to the development time of equivalent products 5 years ago. Also, manufacturers of basically standard equipment such as earth-moving equipment or printing presses win orders by offering well engineered modifications to the product to meet the customers' response.

This demand for rapid response by process planners unfortunately corresponds with a growing shortage of these highly skilled and experienced workers. Traditionally, process planners were recruited from the shop floor, after several years of experience. This route is drying up as fewer and fewer potential process planners are recruited to gain experience by operating machines on the shop floor. The age profile of process planners in British industry is markedly skewed towards the older age groups, so this difficult problem is expected to worsen in the next few years.

There is another problem with process planning as it is often carried out today, that is inconsistency. One would not expect different companies to have identical process plans for identical components, made in the same quantities, because their machines, fixtures and tooling may be different. However, one would expect similar components required in the same volumes to have similar process plans in the same company. Our research shows that for most companies this is not the case. The reason is mainly that it takes too long to retrieve the similar process plan, and it is easier to start from scratch. Of course Group Technology would tend to ensure similarity of process plans, but this may also lead to a problem of slavishly copying a markedly sub-optimal plan.

These considerations led UMIST to start developing computer aided process planning systems. It quickly became apparent that the traditional approach to program structure led to a process planner still having to make the majority of the decisions, as the majority of decisions in process planning, such as feature recognition and machining sequences are not amenable to algorithmic solutions. This led to the investigation of the use of Expert Systems and the realisation that potentially a well applied Expert Process Planning System should be able to deal completely automatically with the majority (say 80%) of components, leaving the remainder to require inputs from a skilled process planner who could also monitor the operation of the system and change process plans if he disagreed with the Expert System. The process planning system produced is called EXCAP.

EXCAP[2][3] is an important element in an intelligent CIM system which is being developed at UMIST. It is intended to manufacture and inspect components automatically from optimum process plans and N-C part programs developed automatically via EXCAP and a tool selection and cutting condition optimising program TECHTURN[1] also developed at UMIST for N-C part program generation.

In parallel with the work on EXCAP for rotational parts, work is well advanced in incorporating automatic feature recognition in ICAPP, which is the UMIST process planning system for 2.1/2D prismatic parts.

All of the work referred to above has been carried out with funding from the ACME Directorate.

Structure of EXCAP

The structure of EXCAP is shown in Fig 1 a to d.

Fig 1a is an over view of the system. The input is a 3D wire frame product model derived from various CAD sources. It is in the GINPUT format. This is fed to an inference engine which recognises features and determines the optimum sequence in which these features should be machined. To do this requires a Knowledge Base of Feature Recognition rules and Feature Machining rules. The knowledge, in the form of POPLOG rules, is entered into the knowledge base interactively by skilled process planners in the particular company in which EXCAP is to be used. This is because a considerable part of the knowledge base is specific to a company because of its particular manufacturing methods and equipment. Self-checking user-friendly interfaces have been written to enable process planners with no knowledge of computer programming to enter rules into the system, to check the validity of rules and to edit them.

The inference engine also requires a database of tools, machines, materials, etc, which is also company-specific. In a similar way this data can be entered and edited via a user-friendly interface.

There are two outputs from the inference engine. One is a hard copy output in the form of a process plan, the other is an output to TECHTURN which creates N-C CL data for transmission to an MHP N-C lathe via a DNC link.

Fig 1b shows the logic of the inference engine. As with most computer based process planning systems, the initial planning starts from the shape of the finished component, determines which should be the last machining operation in the normal sequence, and then "adds metal" to obtain a simpler intermediate state. This reverse process is continued step by step until the bar or forged blank shape is achieved. The inference engine also determines how and where the components should be clamped in the chuck. Some components require machining from both ends and may require both external and internal machining. EXCAP is capable of programming some secondary operations, eg, drilling and tapping a ring of bolt holes, key sinking, and threading, all making use of a power driven tool in the MHP lathe turret.

Once the process plan is completed a colour graphics animation of the machining sequence can be generated if required. See Fig 2c.

Fig 1c shows the various alternative product model inputs to EXCAP. EXCAP is linked to a CV CDS 4001 system and a CV CAD station via IGES 3. This link can be used in principle to interface EXCAP with any CAD system which has an IGES pre-processor. Two additional graphics inputs have been used during the development of EXCAP. One uses a

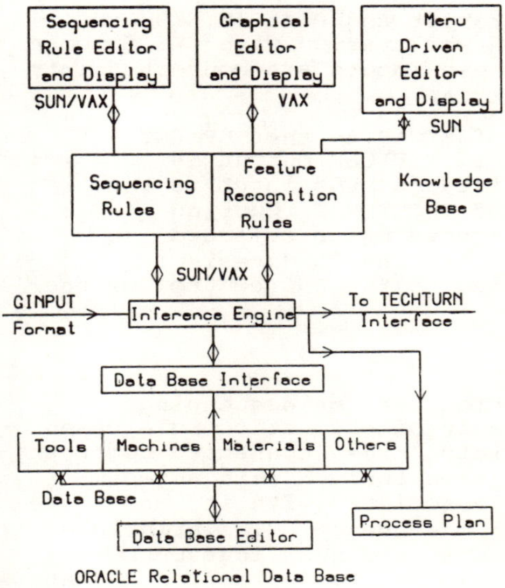

Fig.1a EXCAP

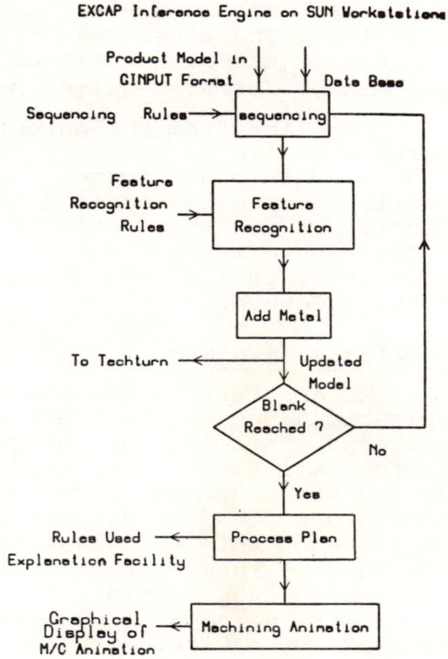

Fig.1b

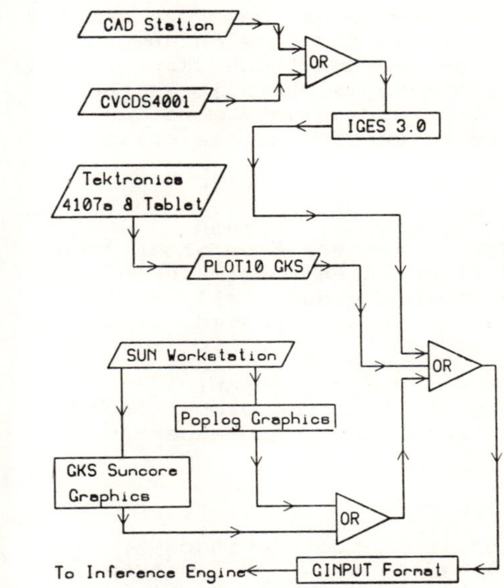

Fig.1c Graphical Input For EXCAP

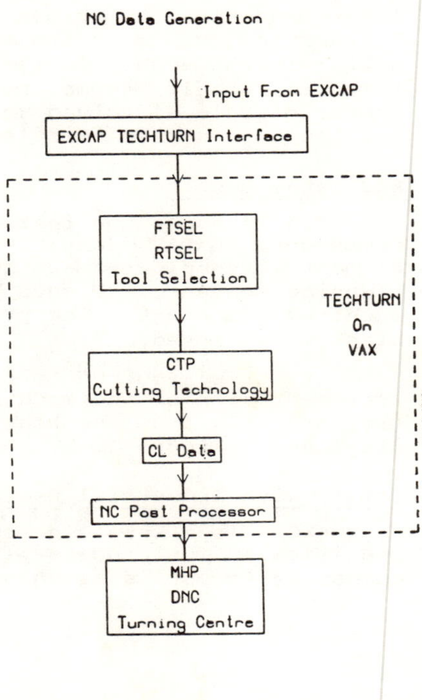

Tektronics 4107a terminal and tablet and Plot 10 GKS graphics. The other uses a SUN workstation with GKS Suncore and Poplog graphics. All of these inputs are converted into the GINPUT format required by EXCAP.

The link from the output of EXCAP to the MHP DNC Turning Centre is shown in Fig 1d. EXCAP output is converted into the format required for the input to the cutting tool selection module of TECHTURN. Cutting conditions are then optimised according to selected optimising criteria, eg, tool life, and CL data is generated, post processed and fed to the DNC control of the Turning Centre.

Knowledge-Base Maps

EXCAP S uses a discrimination net for sequencing operations due to the inefficiency of production rule based systems. This is a set of decision trees connected together to form a network. The route taken through this network during planning has operations associated with it, and these operations form the manufacturing sequence. The principle benefits of this technique are that the ramifications of editing the knowledge-base are explicitly defined, and that the system uses considerable amounts of heuristic knowledge to constrain the search space more efficiently than would be possible otherwise.

There is extra effort involved in editing the knowledge, but due to the explicit nature of the links between rules, it is simple to draw maps of the knowledge-base in which the rules are represented as nodes, and their interconnections as arrows. This technique has been applied hierarchically, to create a 'zoom in' facility on any rule-set. Without such a facility, as the knowledge base grew, it would rapidly become impossible to maintain it as the number of rules involved made it progressively more difficult to access existing rules and to add new rules.

Add Metal Routine

Fig 2a shows the features of a simple turned component. EXCAP identifies that the area from about 3 to element 10 should be machined first in the backwards planning sequence. Figure 2b shows the new partly un-machined component. The process is then repeated until the blank is achieved.

If it is required to obtain a dynamic graphics representation of the machining steps in the proper order (ie, starting from the blank) these can be subsequently displayed. Figure 3.

Trials with Industrial Collaborators

EXCAP is a knowledge engineering tool designed specifically for process-planning and to allow different companies to use is as an expert system shell, capturing the

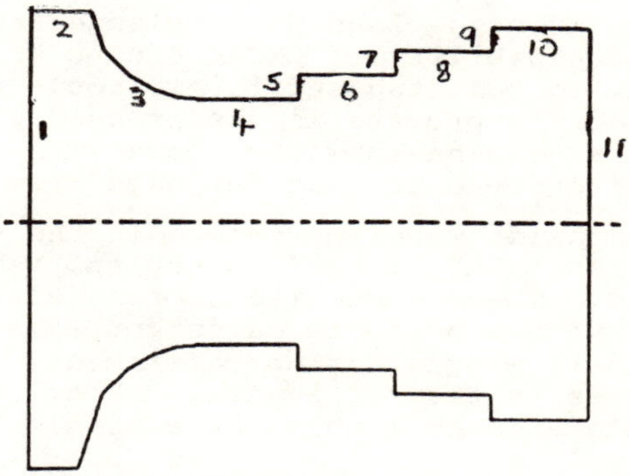

Fig.2a
more-worked workpiece

COMPONENT DESCRIPTION

ID	TYPE	LHD	RHD	LEN	SLEN	OTHERS
1	face	0	10.0	0	0	-
2	diam	10.0	10.0	1.0	0	-
3	blend	10.0	5.0	2.5	1.0	r = 2.5
4	diam	5.0	5.0	2.0	3.5	-
5	face	5.0	6.0	0	5.5	-
6	diam	6.0	6.0	2.0	5.5	-
7	face	6.0	7.0	0	7.5	-
8	diam	7.0	7.0	2.0	7.5	-
9	face	7.0	8.0	0	9.5	-
10	diam	8.0	8.0	2.0	9.5	-
11	face	8.0	0	0	11.5	-

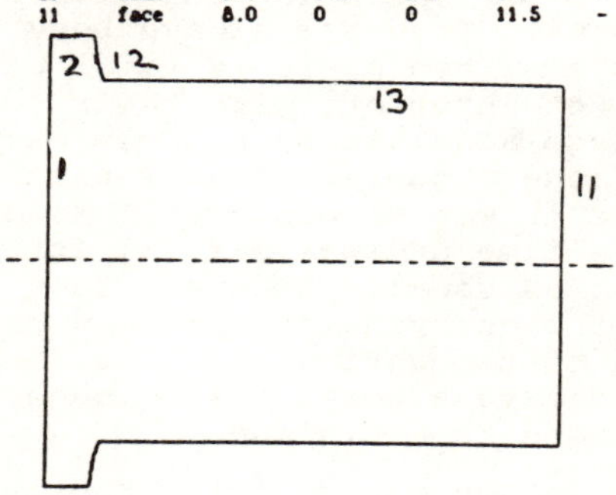

Fig.2b
less-worked workpiece

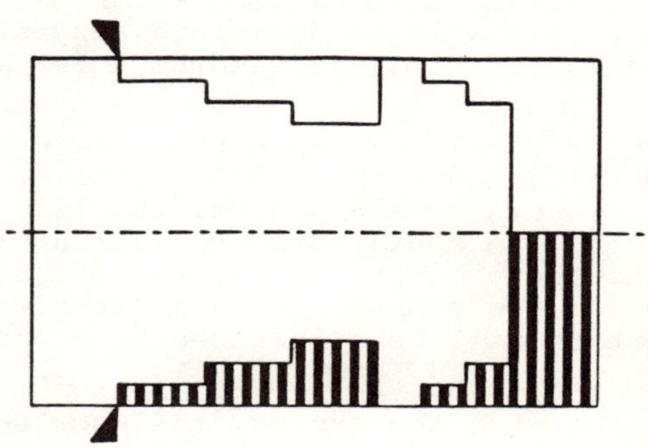

Fig.3
Machining
Animation

company-specific knowledge in the knowledge base. It is necessary to acquire process-planning knowledge from Industry so that the capability of EXCAP can be critically assessed with a view to building sufficient tools to ease the knowledge engineering process in the process-planning domain. EXCAP with a genuine knowledge base will also serve as a demonstration prototype to gain feedback from users.

A number of companies were approached in the early stages of Prolog-based EXCAP, including Leyland Vehicles and Mather and Platt Ltd. These companies cooperated by sending planning sheets, drawings and route-cards for some of their parts, but due to their manufacturing commitments were unable to offer access to their planners, making it impossible to formulate general rules or a usable knowledge base.

A project for acquiring knowledge from process-planners was started in May 1988 with Mirrlees Blackstone Ltd, producers of large diesel engines, as the industrial collaborator. A knowledge-engineer has been working with a planner for three days per month to build up a prototype knowledge base.

Several problems were encountered during the knowledge acquisition process, the major one being the limited availability of the experts. They are always busy and attending to urgent problems. hence, it is vital that the knowledge engineer does not interfere with their job. The major part of the knowledge acquisition process was in the form of shadowing the planner, querying the rationale behind his decision-making while it was happening. This did not take much of the expert's time and was natural for him as he was carrying out his task as usual. However, the interruptions caused by the querying process may potentially break the expert's train of thought, so ideally the knowledge engineer should have a working knowledge of the domain to keep interruptions to a minimum.

Shadowing is a slow method of acquiring knowledge and may not be comprehensive, so the process-planners were interviewed as opportunities arose. These interviews probed deeply the criteria involved in decisions which had been made. Experts were rarely able to formulate general rules, hence 'what if' type questioning was used to infer general rules.

There is a lack of consensus among the domain experts about how a problem in that domain should be solved. This is clear in process-planning where ten experts will give at least eleven solutions. To avoid this problem, at least in the initial stages of the project, an experienced process-planner was selected as the expert for this exercise. The prototype knowledge base will be criticised by other planners later.

After each session with the expert, the knowledge engineer formulated the knowledge into EXCAP routing (sequencing) rules. Periodically, these were shown to the expert and refinements or revisions were made as necessary. This minimised misunderstandings between knowledge engineer and domain expert.

To keep the enthusiasm of the participating expert, the objectives were explained to him in detail and whenever possible the updated software was demonstrated to him. The motivation of the expert was high as he welcomed any aid in his job which would relieve him of the pressure of urgent jobs.

In parallel, UMIST and Rolls-Royce PLC have set up a Teaching-Company scheme. This is a three-year project to apply the techniques and tools, developed in EXCAP, to a very demanding manufacturing environment. Rolls-Royce have a CIM initiative which ultimately requires the automation of a large fraction of the process-planning function and have selected the EXCAP approach as being likely to satisfy their requirements in the mid-term.

Future Plans

A working prototype is being developed with the knowledge elicited from Mirrlees Blackstone. This will be completed before Christmas and will then be tested by Mirrlees by assessing the performance against a panel of process-planners and refining the knowledge base to reduce conflict with the panel.

In parallel, other collaborators are needed with the aim of widening the EXCAP knowledge base. ICL are presently assessing the possible benefits of the EXCAP technology to the manufacturing operations.

The development of an integrated system will continue by linking EXCAP into CAM using technology produced under the TECHTURN project at UMIST, which produces cutting parameters and N-C tape generation.

References

[1] Chan S J and Hinduja S
"Checking for Tool Collisions in Turning"
Vol 20 No 5 Butterworth & Co
June 1988.

[2] Wright A J I L Darbyshire M W Park and B J Davies
"EXCAP and ICAPP: Integrated Knowledge-Based Systems for Process-Planning Components", 19th CIRP Int Seminar on Manufacturing Systems Penn State, June 1987.

[3] Zhang K F, A J Wright and B J Davies, 1988
"A Feature Recognition Knowledge Base for Process-Planning Rotational Components"
Int J of Advanced Manufacturing Technology in press.

Chapter 10

Machine recognition of engineering drawings

D. G. Elliman

1. INTRODUCTION

The term machine recognition describes the process of capturing an image and interpreting its meaning. The input to this process is a drawing on a suitable medium such as paper or microfiche, and the output is a machine readable file.

The introduction of the low-cost laser printer, and page description languages such as postscript (1) has resulted in an explosion of products in the desk-top publishing arena. Illustrations and photographs are often scanned into a DTP program, and this has resulted in a plethora of low cost instruments capable of scanning small drawings. The new generation of photocopiers comprise a scanner, processor, and laser-printer. The image is thus held in computer memory, allowing sophisticated processing before printing. All this has resulted in a vastly increased market for scanners, and low cost devices for A2 drawings will soon be commonplace.

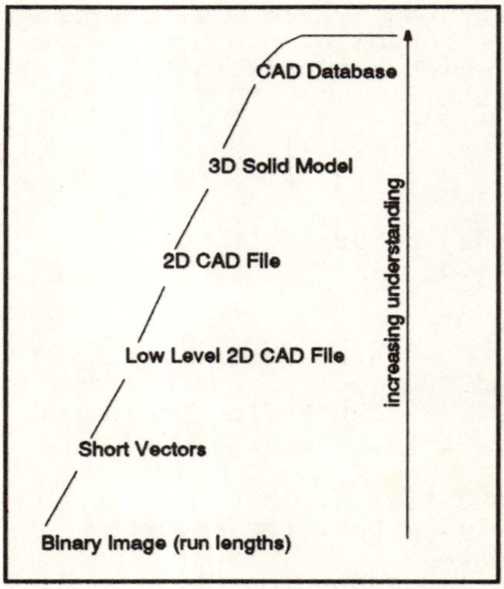

Fig.1 Climbing to Higher Level Representations

The recognition process can be carried to various levels of sophistication as shown in fig. 1. Most existing products merely capture the image, and store it in a compressed form using run lengths. The resulting files are very large, but the recent advances in optical disk and CD-ROM technology make this approach viable for some applications. The drawing can be re-generated at any time using a Laser Printer or Versatec plotter,

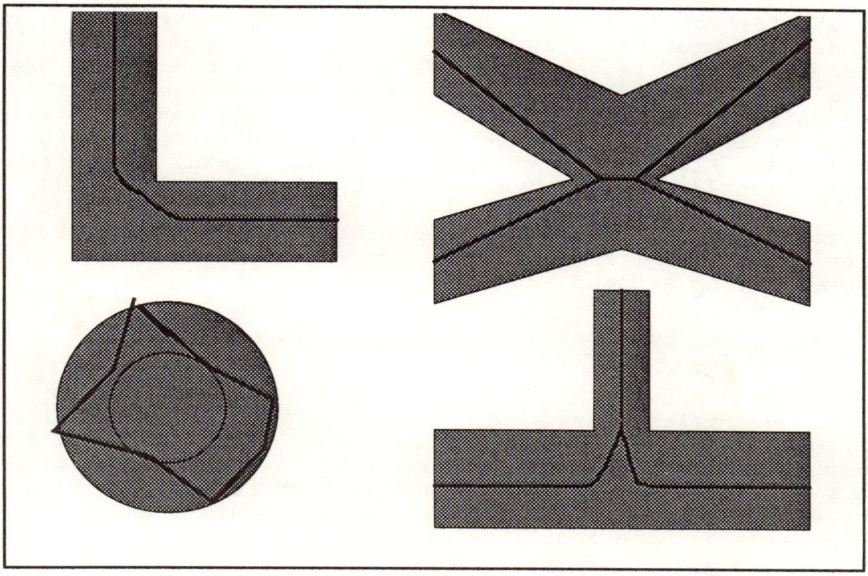

Fig.2 Characteristic Distortions after Vectorisation

and may be transmitted using a facsimile machine. The image can then be used as a background when using a conventional computer-aided draughting program, allowing new features to be added. Such systems are used mostly by public utilities, and local councils who desperately need Geographical Information Systems (GIS), but are frustrated by the lack of digital map information.

Image data is laborious to edit, as this has to be done at the *pixel* level, and occupies an unreasonable amount of storage space. Engineering drawings consist essentially of linework, and it is natural to attempt to transform the image to a *vector* representation. The resulting file then contains many very short vectors, as the original straight lines and smooth curves are now distorted by noise. If this can be removed, the drawing is then represented in a compact, and relatively easy to edit format. A number of systems now on the market claim to be able to vectorise a drawing. The best systems are extremely expensive (over £100,000), and the standard of vectorisation is often very poor. There is usually characteristic distortion at junctions between lines, and in the region of corners and arcs as shown in figure 2. These distortions are a result of the thinning process used to find the medial skeleton of lines on the drawing. The vectorisation process may also be very slow, and in some cases needs continuous manual intervention.

This paper describes a new approach which is able to vectorise drawings very rapidly, and which is free from these forms of distortion. The method described is also able to recognize text, and identify line types on the drawing. This allows powerful noise removal and beautification techniques to be used, and results in a high level CAD file which will reproduce as a higher quality drawing than the original. All arrow heads will be neatly drawn, and characters plotted using a standard fount, for instance.

Further work is under way to correct dimensional errors on the drawing, and to reconstruct a solid model from the views given. Some manual assistance will usually be necessary to accomplish these tasks.

2. THE RAPID PROCESSING OF BINARY IMAGES

An A0 drawing scanned at 12 pixels per mm contains over 100 million binary pixels. Any processing method that needs to examine each pixel will inevitably be slow. If the image is help explicitly in memory up to 16 Mbytes of memory are likely to be needed, and most of the processor's time will be spent in shift and masking operations to set, clear, or examine individual bits in a machine word.

Modern scanners will output a binary image in the form of run lengths of dark and light pixels along each scan line. The data compression achieved by this approach means that a lower bandwidth link between scanner and computer may be used, and results in smaller image files. The key to fast processing is to transform the run-lengths directly to a more compact and usable form. The pixel image is then never explicitly held in memory.

Attempts have been made to vectorise directly from the run lengths (2), but this always leads to very poor quality skeletons perpendicular to the scan line direction. Our approach has been to extract all white to black, and black to white outlines as a tree structure of nested loops, as shown in figure 4. This is achieved in a single pass direct from the run length data. The algorithm works very rapidly, often as fast as the scanner can traverse the paper*

---

* A. further application of this technology is in the field of desk-top publishing. The space occupied by the usual .IMG, and .TIF files used to store scanned images is very large. The outline format used here is much more compact, and can be translated to embedded Postscript without significant space penalty. An image in this form can then be scaled by arbitrary and non-uniform factors. The image compression is greater than that offered by conventional Huffman encoding, and can thus offers saving on telephone transmission costs, and offers a dramatic increase in the speed of transmission to a laser printer. There is a clear market for these capabilities, but no attempt has yet been made to exploit it.

3. MANUFACTURE FROM SCANNED OUTLINES

The loop structure is of direct use for 2D manufacturing operations. For cutting table applications all dark to light boundaries are discarded, and an NC part program derived directly from the coordinates of the remaining loops. The loops would usually be processed inside first, as material is often clamped at the edges.

It is desirable to remove noise from the raw outlines before generating an NC program. Imperfections in the original image, and quantization effects in the scanner result in *wiggly* lines, which lead to a ragged outline and excessively long part program. This is not a trivial problem, and the best solution is application dependent. Outlines are stored as a modified form of Freeman chain (3), as this is compact, and various properties of an outline can be derived cheaply using simple algorithms. An efficient method for eliminating the majority of *wiggles* without loss of information is to find characteristic sub-strings in the Freeman code, and to replace these with simpler but equivalent codes. This is equivalent to the problem of string matching in an editor search, and an efficient method was developed in by Boyer and Moore (4).

Where the outline is known to contain a number of straight lines, an efficient approach is to identify corners, and to attempt to fit straight lines between them. Where this fails, the most distant point is taken as a new corner, and an attempt made to fit two lines. This process of recursive subdivision continues until a line is successfully fitted. Circular arcs can now be fitted through regions of the outline that exhibit regular curvature. The method used simulates the designer's use of french curves, and gives good results without the heavy computational overhead of finding a truly optimum arc. Where straight lines are almost tangential to an adjacent arc, an adjustment is made to ensure a smooth transition of gradient.Further beautification is possible using the methods described by Pavlidis (5), or in simple cases by snapping to a grid.

These techniques are now well proven and ripe for exploitation. They resulted in a reduction in the time taken to generate milling paths for complex artwork from over a week to a few minutes for a sign maker. The computer generated paths were as good as, or better than the previous manual designs. Obvious areas of application include:

- Cutting direct from the pattern (garment industry)
- Cutting fibre cloth for composite lay-up
- Engraving
- Flame cutting
- Defining robot welding paths

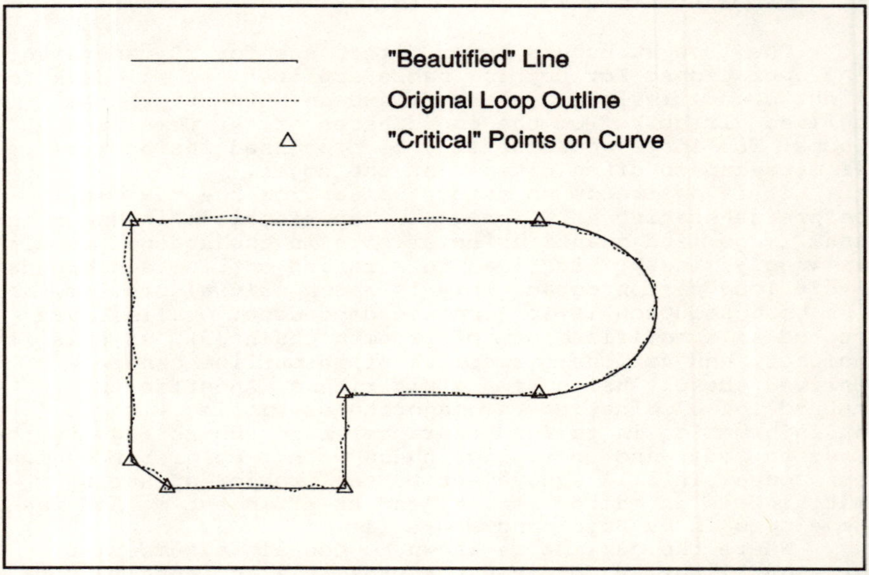

Fig.3 Beautification of "noisy" outlines

- Computer-aided embroidery
- Sign-making
- Profiling turned sections

4. EXTRACTING LINEWORK

To recover linework, each part of a dark to light outline must be matched with a corresponding light to dark one. It is then straightforward to interpolate a median line except where complex junctions occur. In such cases a rule-based approach is used to produce a result that is intuitively reasonable. The rules need to be rather complex to handle every situation, but most junctions are correctly interpolated using the following:

i. Connect any near co-linear lines together.
ii. Extrapolate lines to intersect with lines from (i) provided the point of intersection is reasonable.
iii. Snap together any close points of intersection by sliding along a common line.

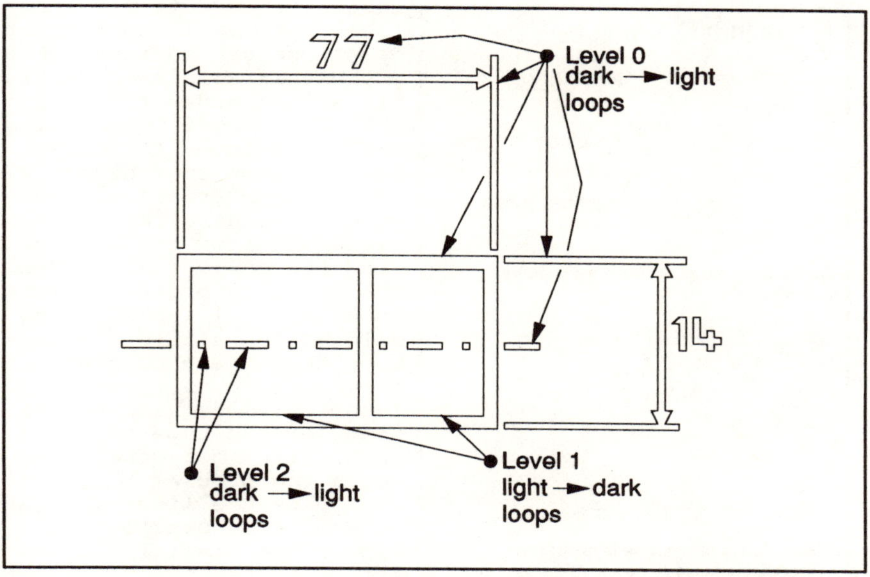

Fig.4 The Loops Structure of a Drawing

Further rules are needed to cover the cases where lines cannot be connected by steps i or ii, and where the intersections are not fully inter-connected.

5. GENERATING A CAD FILE

In order to generate a CAD file at an equivalent level to that produced by manual entry it is necessary to identify the line *types* on the drawing, and to recognize the text strings. The nested *loop* structure provides a number of strong clues to line type, and facilitates the segmentation of text and shading from other linework. Text is always at the lowest level of the nested loops structure, and comprises loops of a tightly constrained range of sizes. The overall outline of a view matches with internal loops, and internal features match with loops at the same level of nesting. In contrast detached linework such as dimensioning and centre lines match within a single loop. These clues, together with identified arrow-heads, and the average thickness of lines form a firm basis for identifying the line types on the drawing. Each line is assigned a set of possible labels on the basis of these clues. Where lines meet (at junctions), constraints can be applied limiting the combination of legal line labels at this point. The set of junctions is repeatedly examined, and impossible

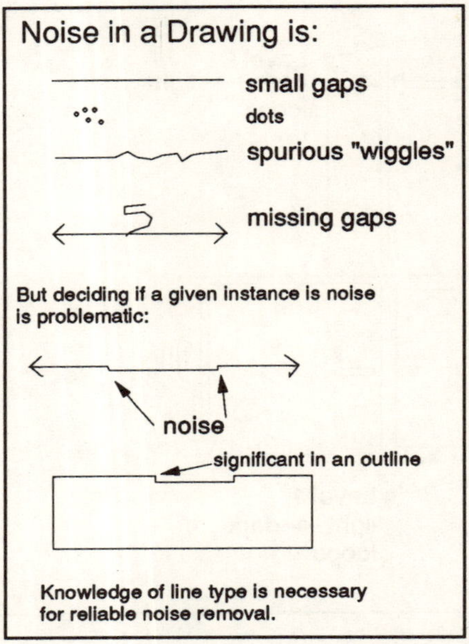

Fig. 5 Noise on a Drawing

labels deleted from a lines label set, until no further deletions can be made. This process is a form of Waltz filtering as described in (6). The algorithm was originally designed for data that is free from errors. In our case, mistakes are sometimes made as a result of mistaken arrow heads, poorly drawn lines, and noise from the scanning process. These result in all labels being deleted for a line, or in more than one consistent interpretation. A rapid interactive request to the user for the type of a single line can often resolve the problem.

Once the line types are known, it is possible to remove the remaining noise from the drawing. Dimension lines and witness lines do not contain small *wiggles*, and will be plotted with the standard arrow head style. Centre lines and dotted lines can be stored as such, and will be drawn with perfect length and spacing. Shaded areas will be held as a perimeter, angle and pitch, and text will be drawn to a standard fount. Small dots and isolated unrecognised marks on the image can then be discarded. The outlines can be beautified using the methods described in section 3.

As a result of this research it is possible to scan existing drawings and convert them into CAD files that differ little from the results of manual entry. This will be attractive to firms which are now using CAD systems, but have a large number of existing drawings on paper. The process should not be expensive in relation to a draughtsman's time, and it may be more efficient to continue to use traditional tools for some drawings, and then to capture them by scanning. The process could also be adapted for the digitisation of maps, offering compact storage forms for geographic information systems.

6. ONGOING RESEARCH

Within the next year it is hoped to achieve a higher level and more structured representation of the drawing. The aim is to understand the drawing much as an expert designer would. Our strategy is to store the labelled lines, and recognized text strings in a layered blackboard structure (7). Various knowledge sources are used to interpret information in the blackboard, and build higher level entities from it. This is a flexible problem solving strategy which has performed well in speech recognitions systems. Each knowledge source is invoked following a cue driven perception cycle. This is illustrated in figure 6. Certain items of data are strong indicators of the presence of a higher level structure. For instance, the arrow head meeting a witness line at right-angles, is a cue for the dimensioning entry shown in the diagram. The knowledge source embodies a model of this feature and tries to instantiate its component lines in the blackboard. If this is successful, the new feature is stored in a higher level of the blackboard, and linked to its constituent lines and text.

Recognized text strings are powerful cues, and knowledge sources are invoked as a result of dictionary look-up. Following this procedure it is possible to gain a high level understanding of the structure and dimensions of each view. For some relatively simple geometries this will be adequate to generate CNC machining instructions, for instance turning, through drilling, and milling at a constant depth. Very worthwhile productivity gains will be forthcoming for particular processes and industries.

7. SOLID MODELLING

For general multi-axis milling, and to gain a true understanding of the drawing, it is necessary to reconstruct a solid model of the component. Known algorithms exist for shapes made from flat faces, but this is a difficult problem for shapes with curved surfaces. Drawings are rarely complete and unambiguous representations of a solid, and an expert system approach is needed to cater for this. Other textual and tolerance information must be understood to represent the component in full. Material, surface finish, and machining instructions must all be captured. The Nottingham group are tackling these problems in collaboration with a team at Trent Polytechnic. Clearly this stage is the most ambitious, and our work is not yet ripe for exploitation. We are confident, however, of producing a working system applicable to certain industries within two years.

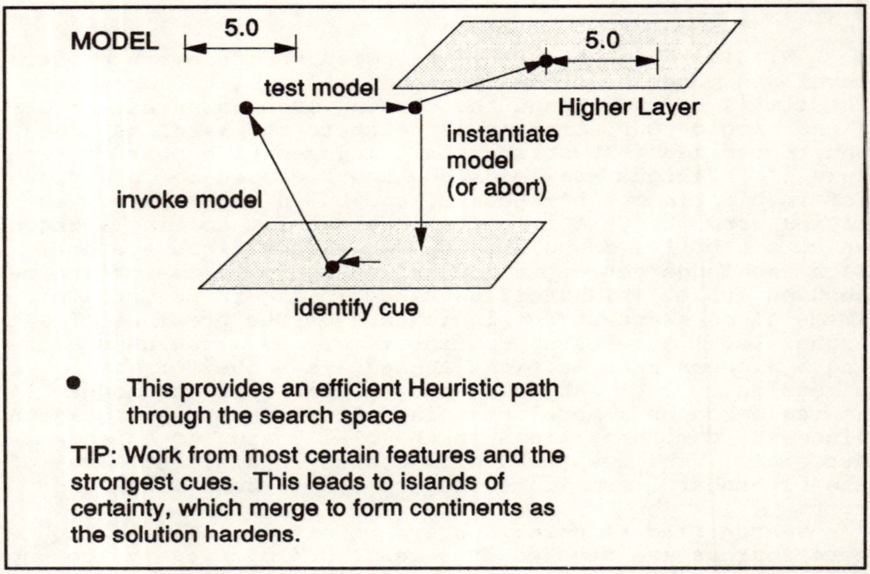

Fig.6 The Cue Driven Perception Cycle

8. CONCLUSIONS

As a result of this research we are able to:

- Rapidly scan images and store them in a highly compressed form. This has many potential applications including DTP.
- Produce 2D machining profiles direct from the pattern. This offers major productivity benefits in various industries associated with cutting, filling, or welding 2D shapes.
- Generate CAD files from a scanned drawing that are equivalent to those from manual digitisation.
- Digitise maps in a form suitable for digital terrain models.

With further work we hope to be able to:

- Construct a solid model from views on a drawing.
- Create a full component database entry from a drawing
- Generate multi-axis CNC miller instructions for complex geometries from conventional drawings. (with some manual assistance)

REFERENCES

1. Adobe systems Inc., 1985, "The Postscript Reference Manual", Addison-Wesley, New York.

2. Pavlidis, T., 1982, "Algorithms for Graphics and Image Processing", Computer Science, Rockville, Md.

3. Freeman, H., June 1961, "On the Encoding of Arbitrary Geometric Configurations", IRE Trans. Electronic Computers, 260-268

4. Wirth, N., Sept. 1984, "Data Structures and Algorithms", Scientific American, Vol. 251, No. 3, 48-57.

5. Pavlidis, T., July 22-25 1985, "An Automatic Beautifier for Drawings and Illustrations", Proc. SIGGRAPH, San Francisco, 225-234

6. Waltz, D., 1975, "Understanding Line Drawings of Scenes with Shadows", McGraw-Hill, New York.

7. Engelmore, R., Morgan, T., 1988, "Blackboard Systems", Addison-Wesley.

ACKNOWLEDGMENTS

The support of the ACME directorate of the Science and Engineering Research Council for this research is gratefully acknowledged.

Chapter 11

A knowledge-based system for fatigue design

C. A. McMahon, J. H. Sims Williams and J. Devlukia

1. INTRODUCTION

The Departments of Mechanical Engineering and of Engineering Mathematics at the University of Bristol are collaborating with the Gaydon Proving Ground of the Rover Group on the development of a knowledge-based system for the design and development of automotive parts. The project is supported by the ACME directorate of the Science and Engineering Research Council.

The project is currently nearing the end of the first year of a three-year programme. The target for this programme is to combine the use of database and expert systems techniques to provide support for the designer in design for optimum fatigue performance. This paper is intended to:

Overview the present automotive design methodology, and the methodology that we propose to allow knowledge-based systems to be exploited;
Describe the structure of the system that is being constructed;
Report on our initial progress towards implementation of the system.

2. DESIGN METHODOLOGIES

2.1 Present Methodology

The current design-to-product cycle for automotive components frequently follows an iterative loop of design, prototype manufacture, test and redesign. In this loop four key elements can be identified:

Design layout and detail design;
For complex or critical components, a design analysis phase may be included;
Prototype manufacture and test;
In the event of a failure of the component to achieve satisfactory durability or performance, design modifications will be carried out. The modifications will be guided by the results of prototype test, and perhaps by extensive analysis.

The third and fourth elements are repeated until a satisfactory design is achieved. The whole process has a major emphasis on development testing, and on incremental modification of existing designs.

The use of this iterative approach is expensive in time and resources, and in particular results in long design/development lead times. The approach also tends to produce designs which are developed to function satisfactorily, rather than optimally. There is strong pressure in the automotive industry for both lead times and product quality to be improved.

2.2 An Improved Design-to-product Cycle

We suggest that there are two targets that should be achieved in order to improve the design-to-product cycle. Firstly, the initial design phase should produce designs that are more nearly optimal. Secondly, design analysis should be applied as a matter of routine at the initial design phase for a wider range of components. If these two targets are achieved, then the development effort will be freed to concentrate on detail design improvement, verification and optimisation.

In order to achieve the first target, the designer must be provided with an alternative to intuitive or experience-based incremental design. We suggest that this alternative should be based on an improved supply of design information to the designer, and on the application of specialist expertise at the initial design phase. At present the supply of design information is not formalised, and the provision of expert advice is generally on an occasional consultative basis only.

The achievement of the second target is limited by the time and expertise required to undertake design analysis, and by specific limitations in the available analytical techniques. While it is acknowledged that detailed analysis of critical components will remain the domain of the specialist, we suggest that the second target can be partly achieved by improving the ease of application of analytical techniques by the designer, and by providing expert assistance in the selection and execution of the analysis.

The design process currently, as described above, leads to a prototype and invests in experimental testing at an early stage in the iterative process of design improvement. The basis of the proposed knowledge-based method is to make heuristic and analytical tools, AND expert advice in their use, available to the designer at a pre-prototype stage. These tools will be directed at providing quick estimates of the properties of designs, rather than providing definitive analyses. In this way the designer will be able to go around the design-analysis cycle several times in quick succession.

At present, methods available for data retrieval are often laborious, and many designers do not have the necessary expertise to use modern analytical methods. We propose to give the designer access to a computer-based integrated design aid that will combine an expert-system for rule based advice, a database system for storage and presentation of engineering design data, and facilities for the execution of procedural or algorithmic tasks. The expert system will lead the designer through the procedures, and provide him with the tools necessary for decision making and analysis.

Rather than attempt in the first instance to provide an all-encompassing design aid, we propose to test the methodology using the vehicle of design for fatigue. This subject has been selected for two reasons. Firstly, it is estimated that more than 80% of failures arise from fatigue and therefore this is a significant problem in itself for designers. Secondly, both the Gaydon Proving Ground and the University have considerable expertise in this field.

3. THE KNOWLEDGE BASED SYSTEM

3.1 Expert System Technology

Expert systems are computer programmes that are distinguished from normal programmes by separation of the knowledge or expertise from the 'inference engine'. The inference engine is the part of the program that searches through the 'knowledge-base' in response to external signals such as user queries, and returns answers or recommendations.

The separation of the knowledge-base from the inference engine has several advantages:

The inference engine can be used for several different applications, enabling new applications to be generated rapidly;

It is possible to have a knowledge-base written in an English- like form so that:

(a) Non-programmers can understand the knowledge and thus be able to correct and improve it, as noted by Davis and Lenat (1).

(b) Reference can be made to the knowledge-base to give explanations of the advice given by the system;

In a conventional program, the knowledge is often intermingled with the program code, and so a programmer is required to change the code when the knowledge needs updating. This considerably increases the cost of maintaining the software.

One of the techniques for writing expert systems is in terms of objects which can be real objects such as machines or components, or can be concepts such as productivity or pollution. These objects can be arranged in hierarchies so that properties are inherited from super-objects in the same way that we understand the world in terms of classes of objects and ideas. These objects are then given properties and ways of doing things (methods) so that the expert system effectively simulates the way in which all the objects will interact in order to make its recommendations. Sometimes this object oriented way of thinking fits in closely with the way engineers conceive their problems and then it reduces the problems of explaining the expert systems recommendations and updating the knowledge.

Engineering design knowledge is vast in extent and is always changing. To match up to the task of capturing, let alone maintaining, this body of knowledge it is essential that the designers and analysts themselves are able to generate and update the knowledgebase. If designers are to do this task they must understand why the system makes particular recommendations, so an explanation facility should be provided, and the knowledge must be represented in a way that is natural for designers. Figure 1 illustrates the way in which software systems are provided in order to help both with the explanation of recommendations and the acquisition of new knowledge.

3.2 Database Technology

The whole process of design is very heavily dependent on data. There will be data on previous designs, their successes and failures, data on the materials that can be used, on the load lives of components and so on. A significant proportion of a designer's time is spent on looking for data and keeping it current. A database management system (DBMS) is designed for this purpose. We propose to make databases available to the designer at his workstation, so that he can peruse displays of the data. The expert systems will also need access to the same data to provide advice and assistance.

At present relational databases represent the most general and well-organised way of storing data. They store data in the convenient form of tables of rows and columns. One can however have a wide variety of elements in the tables, such as bit-mapped images (eg of photographs), vector representations of drawings, graphs, rules for expert systems, numeric data.

A new technology is developing called 'Object-Oriented Databases'. It is expected that these new systems may have advantages when using object-oriented methods for expert-systems.

3.3 Overview of the Proposed System.

The essential form of the proposed knowledge-based system will be a program that executes in one or more windows of an engineering workstation, and which provides a series of tools to the designer. These tools may be accessed individually in a random manner, or the designer may be guided through their use. The tools will be divided into:

3.3.1 A fatigue life assessment expert (FLAE). The aim of this part of the system will be to allow the designer to predict, for a given component geometry, material and loading, where the component is likely to fail in fatigue, and what the component life is likely to be. In order to achieve this, the module will provide advice on:

selection of the appropriate fatigue analysis model for a given component (eg local strain, S-N, fracture mechanics or other approach, multi-axial model and so on);
selection of the appropriate load history and service life of the component;
identification of the critical locations of the component, and on the appropriate stress concentration factors and notch sensitivity factors for these locations;
preparation of input data for a variety of fatigue analysis models;
interpretation of the results from analysis;

In addition, various fatigue analysis algorithms will be built into the system.

3.3.2 A fatigue life modification expert (FLME). When the user has assessed the fatigue life of the design and compared this with requirements, he may wish to increase the fatigue strength of the component or even decrease it for economic reasons. The FLME will provide advice on how to modify the fatigue life of a design by changing geometry, material or method of manufacture. This tool could also be used at any time to advise on design improvements for existing components, or subsequent to development work.

3.3.3 A database system. Each of the tools described above will be dependent on data on the properties of materials, the effects of surface treatments and manufacturing processes and the sort of service duty that the component is expected to survive. The expert systems will be built around a database system to store this information.

4. PROGRESS TOWARDS IMPLEMENTATION

The project commenced at the beginning of 1988. Effort has been concentrated in the first stage of the work on:

identification of computational tools for the systems, and on an overall expert system/database architecture;
preparation of a requirements specification, and knowledge elicitation based on this specification;
the construction of small prototype systems to address key elements of the problem. In particular, the

questions of representation of geometry, and of object-oriented approaches, are being tackled.

This section describes in more detail the experience gained on this work.

4.1 Choice of Expert System Tools

Expert system technology can provide the opportunity for rapid prototyping of systems. This is possible because of the separation of the inference procedures from the knowledge. It is possible to purchase Expert System Shells in which the inference procedures and often the user interface are provided so that it is only necessary to add a knowledge base in order to build a prototype expert system.

Examples of commercial shells are Crystal (2), Xi and Savoir. In addition to the shell approach, the system developer may also choose to use AI languages such as LISP, FRIL, Prolog and Smalltalk, or expert system toolkits such as KEE, Inference ART or Nexpert.

Shells allow one to develop a prototype extremely quickly, but they tend to allow only a single mode of representation of knowledge and the user-interface is often not very versatile.

AI languages allow enormous freedom to use any mode of inference and knowledge representation but mean that development times are much greater. Toolkits allow one to use a defined variety of knowledge representation techniques and allow an interface to other languages for both user-interface and computational purposes. This versatility is costly, but we expect that the development time savings of using a toolkit compared with an AI language will justify the extra cost. We have decided to use 'C' as the language of implementation and chose a toolkit (Nexpert) which was written in 'C' to help us with the rapid prototyping of expert systems.

4.2 The Design of the System

We have identified that designers using our system, which we shall call a Designer's Assistant, will require to have access to considerable volumes of data in the form of tables, graphs and text; they will require access to various expert systems which will in turn need access to large volumes of data such as material properties, stress concentration factors and load life descriptions. Figure 2 illustrates the simplest possible structure for the Designer's Assistant. During the development of a design, however, data will continue to be added and so, for the efficient operation of the system, we need to have a data controller which will keep track of what data is known and stored in one of the databases, and will ask the designer for other data as it is requested by one of the expert systems. Al Zobaidie and Grimson (3) have discussed the different ways in which control is best introduced, and we have adopted their system, called Difead.

The design of the system, based on Difead, is shown in Figure 3. The user talks through a user interface direct to the controller which handles databases (DBs) and expert systems. This has the advantage that a separate user interface is not required for each expert system, and it improves the standardisation of interfaces. The databases can be updated by their own management systems, so it would be possible to have a company database as one of the databases on the system.

The controller has a data update module which allows it, or the user, to update the databases using information from either the user or an expert system. In addition, the controller has two data dictionaries (DD). The Difead DD contains simple data on the system-primarily which databases exist and which expert systems. The Control DD contains detailed information for each expert system about the location of all the knowledge and data that the expert system may require. The Control DD makes sure that data are retrieved and stored efficiently.

The user of the system may require to run programs for his own benefit - eg S-N fatigue calculations - or those programs may be run by an expert system. Programs are thus treated by the controller in the same way as databases. At present we are planning to run the Designer's Assistant in a window of the workstation with a CAD package and Finite Element program run separately in different windows. This weak link to the controller is depicted by the dotted lines in Figure 3.

4.3 Representation of Geometry.

The identification of the locations on a component that are likely to be initiation sites for fatigue, and the subsequent analysis of the component, involve an interpretation of the interaction between the component geometry, loads, material and surface condition. The latter two aspects are relatively straightforward to deal with in a computer program, but the description of the geometry and load case is more problematic. There is considerable research being undertaken at present into the identification of component features from Computer-aided Design models of the geometry as for example described by Woodwork (4), and this may ultimately identify a method that may be used in a fatigue analysis system. In the short/medium term, however, it is considered that automatic processing of CAD models will not be fruitful because:

many automotive components are complex castings or forgings that are not (readily) described completely using current generation CAD systems. It will be necessary to be able to use conventional drawing representations for some time to come;

many component features that are likely to be fatigue initiation sites do not conform precisely to standard models - in other words there is a considerable amount of expertise used in the interpretation of geometry.

An alternative strategy, which has been rejected on the grounds of complexity, is to carry out a detailed finite element analysis of the complete component. Instead, we propose that the user identifies critical features of the component by an interactive method in which he user will be shown a series of pictures of standard stress-raising features, and asked to indicate which features correspond most closely to those on the component. Heuristic methods will be used to convert the known features of forgings, sheet metal details and so on to standard stress-raising features as identified in works on stress-concentration factors such as Peterson (5). Heuristic methods will also be used to reduce the range of possible features for the component (for example, sheet metal details will only exhibit a small range of stress-raising features).

5. CONCLUSIONS

In developing the knowledge-based system, it is not the intention to try and replace human experts in fatigue analysis, but rather to allow the more mundane and routine analyses to be undertaken by non-specialists, freeing the specialists to deal with special cases. It is hoped that these cases will, in turn, be built into the system and become routine. In order to achieve this objective, many hurdles have to be crossed. Of these, it is considered that the questions of the representation of geometry, and of the structure of systems that may be modified and updated by the user, are paramount.

6. REFERENCES

(1) Davis, R., and Lenat, D, 1982, 'Knowledge-Based Systems in Artificial Intelligence', McGraw Hill.

(2) Firth, P., and Nealon, J., 1988, 'Expert Systems for Process Planning', SERC, ACME Directorate Research Conference Proceedings.

(3) Al-Zobaidie, A., and Grimson, J.B., 1987, 'Expert Systems and Database Systems: how they can serve each other', Expert Systems, 4(1).

(4) Woodwork, J.R., June 1988, 'Shape Models in Computer Integrated Manufacture - a Review', Computer-Aided Engineering Journal, 5(3), 103-112.

(5) Peterson, R.E., 1974, 'Stress Concentration Factors', Wiley Interscience, New York.

FIGURE 3

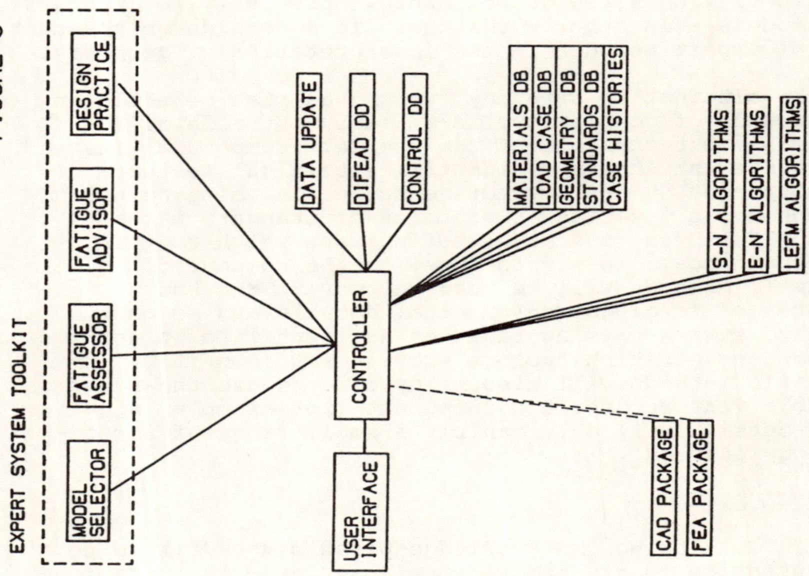

FIGURE 1

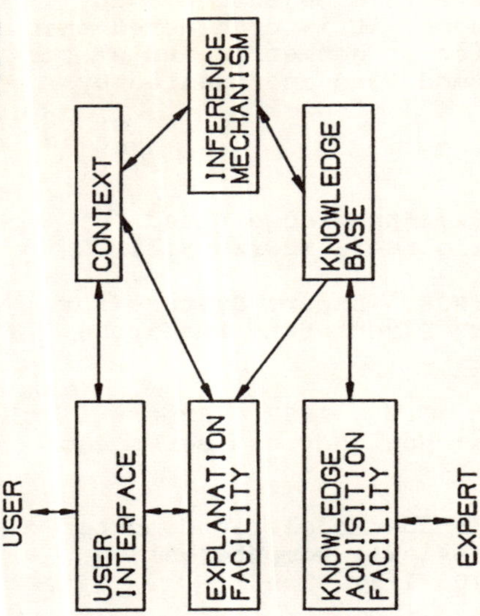

FIGURE 2

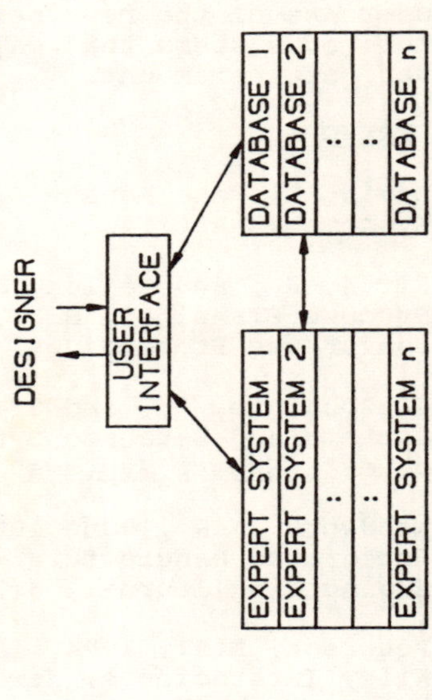

A SIMPLE DESIGN OF THE DESIGNER'S ASSISTANT

Chapter 12

The state of the art of AMT implementation in the ceramics industries

K. Ridgway, I. K. Bloor and S. K. Ghosh

1. INTRODUCTION

This paper discusses the interim results of a study, supported by the ACME Directorate of the SERC and British Ceramic Research Ltd, Stoke-on-Trent, to examine the application of AMT in the Ceramics Industries. The Ceramics Industries are traditional industries incorporating a high degree of manual skills. At first sight it would appear that they are ripe for the implementation of AMT. Yet many companies have been reluctant to invest in AMT as they have yet to be convinced that it is applicable to their process orientated industry.

After an initial description of the Ceramics Industries, the article discusses current AMT activities within the Industries. The specific requirements of the Industries are discussed and several factors inhibiting AMT introduction are identified.

2. THE CERAMICS INDUSTRIES

2.1 General

The Ceramics Industries can be conveniently divided into a number of sectors as shown in Table 2.1. The table indicates the relative sales value and the number of employees in the various sectors. For convenience the sectors identified can be grouped into the Whiteware, Industrial Ceramics and Heavy Clay Industries. The table includes the relevant information for these groupings provided by the Business Statistics Office (2). The Whiteware and Industrial Ceramic Industries are characterised by a diversity of product which leads to short production runs and high inventory levels (typically 40,000 items in the case of a high quality china factory). Lead times are typically 8 weeks and it is estimated that between 30% and 40% of finished ware is recycled due to imperfections identified during inspection. In addition the manufacture of a ceramic product involves a number of discrete steps. This leads to large buffer stocks and high levels of work in progress. This is not the case in the Heavy Clay Industries where the diversity of the product is not as great and a more continuous production process can be utilised.

3. CURRENT APPLICATIONS OF AMT IN THE CERAMICS INDUSTRIES

3.1 General

This section examines the state of the art of AMT in the Ceramics Industries. It discusses several techniques which have been successfully applied in the engineering industries and discusses problems specific to the Ceramics Industries. The techniques examined include CAD/CAM, CAPM, Robotics and Automatic Procss Control.

3.2 Computer Aided Design and Manufacture

The application of CAD is relatively new in the Ceramics Industries although it has been readily accepted in the fringe Industries such as refractory construction. In this Industry 2D and 3D systems with facilities for parts listing have achieved considerable success (Lutcke (32)).

The benefits of introducing CADCAM techniques have been investigated by a number of manufacturers. One study recently reported by Clayton (7),(8),(26) examined the application of CADCAM in the design and production of moulds for the manufacture of sanitaryware. The study carried out by British Ceramic Research Ltd involved nine leading European manufacturers and concluded that a significant economic adavantage could be gained by using CADCAM.

The study used the Deltacam three dimensional modelling package DUCT to generate a model of the desired component. With the assistance of an experienced modeller the model was adjusted to compensate for shrinkage during firing. A plaster block mould was then produced and several prototypes were manufactured. On completion the prototypes were compared to the desired shape and the shrinkage compensation was adjusted as required.

Work currently being carried out at Josiah Wedgwood Ltd by Dillon (27) involves the use of of the DUCT package for initial product visualisation, trial marketing amd mould manufacture for the production of tableware. It is estimated that the use of CADCAM can reduce the lead time from fifteen weeks to less than three weeks.

The use of CADCAM to assist in the design of dust press dies is currently being investigated. A major factor in the cost of dust press dies is the need to undertake a number of iterations to achieve a die which will produce the desired form after firing.

The project which will be carried out by British Ceram Research Ltd (1) will examine the use of a mathematical model to predict the die shape from the fired component. Work carried out to date indicates that the full benefits of

CAD/CAM will only be realised when a number of problems have been solved:
i) Moulds and dies must be designed to allow for shrinkage of the product during firing. A relationship between the CAD target model and the shape of the mould or die must be determined.
ii) The packages must be user friendly and suitable for use by artistically trained staff who are not initially familiar with CAD.

3.3 Computer Aided Production Management

CAPM has been introduced into the industry with varied success. Clayton (9) states that success has been achieved where the software has been modified in house to suit the company's requirements. Several studies (Hooper (10) and Nitchman (11)) have indicated that computer systems could be used for inventory control, sales order processing, purchase order processing and financial accounting. As reported by McLaughlin (31) the only companies to install large scale stock control systems are Armitage Shank's, Rugeley and Coloroll Ceramics, Meir. Both have centralised distribution depots which have involved large capital investments.

The Industries comprise many small/medium sized companies who would by necessity turn to less costly package software. In general, packages are produced for the engineering industry and modifications are often costly. The Ceramics Industries have unique requirements due to the characteristics of the manufacturing processes described earlier. Computer Aided Production Management systems offer considerable benefits provided that a suitable software packages can be identified. Companies which have successfully implemented CAPM have selected systems which are easily adapted to their needs.

Further work is required to specify the detailed requirements of the industry and to identify software packages which can be easily modified to meet individual requirements.

3.4 Robotics

Several studies have examined the application of robotics to the Ceramics Industries. Guha (21) suggests that robots can be used for loading and palletizing pressed components and handling components during measurement checks. He further suggests their use for loading and unloading kiln cars and spray glazing whitewares. These latter applications are also identified by Birch (12) who also includes the handling of delicate components and assemblies. Shivarov et al (25) carried out a study to examine the use of robotics in the palletization of wall tiles. They concluded that the pay back period was 3.5 years but made the proviso that the manufacturing process was completely automated and that the robot operated under optimum conditions. From the above the main areas identified for the use of robotics are;

i) handling fragile components,
ii) loading and palletizing wall tiles,
iii) spray glazing whitewares.

Several companies are making progress in these areas. Selander and Lemke (20) report that IBM who have installed a robotic system for loading and unloading ceramic sheets at an inspection station. The sheets are extremely fragile and a key factor in the success of the robot has been the design of the end effector which picks up the the sheets using a vacuum head. The inspection station has successfully exceeded the target production increase of 28% and the operator requirement has been reduced from four to one. Edward Curran Ltd (24) of Cardiff are currently using robots to spray vitreous enamel onto steel bathtubs which are sold under the Twyfords name. This is a similar technique to glaze spraying and demonstrates that the application is feasible. Several companies particularly tile and refractory manufacturers (33) have installed automatic handling systems. A number of technical problems are discouraging greater use within the industry. These were highlighted by Graf and Nicolaisen (22) who carried out a detailed evaluation of the application of robotics in the Ceramics Industries. They identified the main problems as:

i) Handling: Ceramic components are brittle or highly deformable. The diversity of shape is also a problem particularly when loading kiln cars is considered. Although this is often cited as an ideal application for robotics it is rarely the case due to the diversity of products loaded at any one time.

ii) The production cycle: The layout of many factories is unsuitable for the introduction of robotics. This is particularly evident when visiting sites in the Stoke-on-Trent area where many companies operate in multi-storey factories. In addition handling operations such as decorating are carried out manually for artistic reasons and automated systems would not be applicable.

iii) Economic factors: The level of investment risk accepted by the whitewares industry is low. When investment funds are allocated, priority is given to product improvement.

(The findings of the German study were confirmed by visits made to British manufacturers).

Robotics have been shown to be applicable to the industries in a number of studies but several of these have been based purely on economic considerations and have ignored the practical problems which can occur. Problem areas which have been identified include the design of end effectors for handling fragile and easily deformable components and the suitability of the environment.

Robotic systems have been successfully implemented where care has been taken with gripper design and where the layout of the manufacturing facilities has been favourable.

The tile and brick industries have adopted robotic systems more readily because the nature of the production processes and the shape of the product are more suitable to robotic handling. Several of the previous studies have identified brick and tile applications due to the simplicity and consistency of shape. It is apparent from company visits that many companies would have to replan the production environment before robots could be successfully introduced.

Further work is to be carried out to evaluate the feasibility of installing robotic systems in the production of whitewares. This will consider the logistics implications in addition to a detailed financial analysis.

3.5 Automatic Process Control

The introduction of automatic process control is easier to achieve in continuous high productivity processes such as tile and brick production. Brick manufacturers are using PLCs for the control of a variety of material handling and packaging functions as described by Burchell (34). Computers are also being used for monitoring kiln and dryer temperatures.

In the Whiteware and Industrial Ceramic sectors automatic process control has been introduced for individual processes. A typical example being kiln control where a number of companies are offering computerised systems as described by Leisenberg (13) and Tinsley (19). The facilities offered include automatic control of such variables as temperature in the firing zone, flue gas volume and inlet air pressure. Rafferty (23) describes work which has also been carried out on the automation of kiln car handling systems where a variety of systems for loading kiln cars have been developed.

Thomas (14) and others (28) report that glaze dipping, glaze spraying and fettling processes have also been automated by the introduction of microprocessors in the control systems.

Visits to the manufacturers during the current project indicated that automated weighing systems could be particularly useful. As reported by Clarke (16) Twyfords have introduced microprocessor control of raw material weight and body recipe. This gives more accurate control of the recipe, costing and stock reordering in addition to reducing material handling.

Miotto (15) carried out an extensive study into the automation of tableware production. He identifies linking discrete processes and the limitations due to the inherent shape of the product as major problems. He suggests that research should concentrate on the development of new production techniques which could be more readily automated.

Visual inspection was also identified as an area where modern technology could be introduced. Machine vision systems are now capable of identifying defects in colour, design and shape. This can be achieved using a computer controlled unit which utilises video cameras to compare tiles with reference pieces. One system described by Medau (17) is reported to have been installed at Villeroy and Boch in Mettloch, W.Germany, has achieved 96% agreement with a manual inspection system which is currently used on the same production line. At present the technology is restricted to the tile industry. Inspection of tableware would be a more difficult task and simple vision systems alone would not be able to cope with the intellectual decisions required. Although machine visions systems could detect such defects as kiln dirt, black spots, colour bleeding and chipped edges they would be incapable of judging whether the defect would be detrimental on completion of the manufacturing processes. A black spot or kiln dirt may indicate a defect at the pre-decoration stage but this may be obscured by subsequent decoration. In addition in several sectors of the industry there is an acceptable percentage of defects which can be included in production quotas.

4. CONCLUSIONS

The full benefits of introducing new technology into the industry are rarely achieved. This is due to human issues which are rarely included in the feasibility equation. Guerrieri (18) identifies the changing operator skill requirements and the reluctance of companies to retrain staff as critical factors. The British Ceramics Industries are traditional industries centred in Stoke-on-Trent. The skills involved are unique to the Industries and there may be reluctance to accept new technology. It has been found that designers have been reluctant to use unfriendly CAD at the preliminary design stage.

The visits to ceramics manufacturers indicated that the layout of the equipment and processes could often be improved. The traditional multi-storey factory is not conducive to the introduction of automatic handling systems and much could be done to improve the flow of products through the factory.

A recent study carried out by Chabinsky and Eves (35) indicated that the use of micro-wave energy for drying and firing could increase productivity, reduce costs and improve quality. It is clear that the processes used in ceramics production could be more conducive to AMT implementation.

In the next stage of this study a systems approach will be applied to ceramics manufacture and the benfits of linking processes within a Flexible Manufacturing Cell will be evaluated.

Quality control appears to be a problem when the rejection levels of 30-40% are considered. It is accepted that a market exists for second quality goods but a recent study reported by Kay (29) demonstrated that a considerable improvement could be made by introducing experimental design techniques. Experimental design (more commonly termed Taguchi methods) is used to examine the relationship between product and process variables . It is particularly suited to the manufacture of ceramic products as the optimum selection of process variables has a considerable effect on product quality.

It is clear from this study that there are many reasons why the Ceramics Industries have been reluctant to implement AMT. If AMT techniques can be shown to be applicable and economically feasible the industries will readily accept their introduction. However, the application and feasibility must be demonstrated within the constraints placed on the manufacturers. AMT vendors must accept that the layout and production processes may not be ideal and the economic evaluations must consider these factors.

5. ACKNOWLEDGEMENTS

The authors wish to express their gratitude to Mr W L Walters of BCRL for his assistance throughout the project. They also wish to thank the companies who have participated in the project to date. The ACME Directorate of SERC provided the funding for this initial project and their assistance is gratefully acknowledged.

6. REFERENCES

1. British Ceramic Research Ltd, 1987
   Annual Report

2. Business Statistics Office, 1987
   "Business Monitor, Quarterly Statistics
   (Ceramics Goods PQ 2489)"
   HMSO.

3. Humphreys H, 1988
   "Designing Sanitaryware with Deltacam DUCT"
   Ceramic Industries Journal
   94 (1054) 28

4. Clayton C.G.A. 1987
   "The Sanitaryware Factory of the Future"
   British Ceramic Research Ltd,
   Given at Interceramex,
   Stoke-on -Trent

5. Clayton C.G.A. 1987
"The Tableware Factory of the Future"
British Ceramic Research Ltd,
Given at Interceramex,
Stoke-on-Trent

6. Clayton C.G.A. 1987
"The Wall and Floor Tile Factory of the Future"
British Ceramic Research Ltd.

7. Clayton C.G.A. 1987
"Application of CAD/CAM"
British Ceramic Research Ltd.

8. Clayton C.G.A. 1988
"Application of CAD/CAM to the Production of Whitewares"
British Ceramic Research Ltd,

9. Anon,1985
"Integrated Financial and Production Control System
For Yorkshire Brick".
<u>Ceramic Industries Journal</u>
<u>94</u> (1051) 36.

10. Hooper R. 1985
"Computers in the Ceramics Industry",
<u>Ceramic Industries Journal</u>
<u>94</u> (1054) 26.

11. Nitchman P.P, 1980
"Solving Production Problems with a computer",
Ceramic Engineering Science Proceedings,
1 (7110) 16.

12. Birch M.A, 1979
"Microprocessors and Ceramics: Role of the Robot"
British Ceramic Review,
<u>(37)</u>, 21.

13. Leisenberg W, 1985
"Computerized Control of tunnel kilns",
<u>Ziegelindustrie International</u>,
<u>38</u> (9) 502.

14. Thomas R.J, 1987
"Microprocessors Uses in a Sanitaryware Plant - From simple to complex".
American Ceramic Society Bulletin <u>66</u>
No4.

15. Miotto S, 1987
"Automatic Modules for Crockery Production"
PT1 & PT11,
Interceram 36,
36 (3) 23, (4) 26.

16. Clarke D.G, 1982
"The Potential for Process Control in Sanitaryware"
Trans Journal British Ceramic Soc.
81 (5) 132.

17. Medau H, 1986
"The fully automatic Q C inspection"
Ceramic Forum Int/Ber DKG
63 (4/5) 213.

18. Guerrieri G, 1982
"Automation and Ceramic Industry Problems and Perspectives"
Ceramic Informazione
17 (192) 135.

19. Tinsley D.I, 1987
"Imposed Control of Kilns"
British Ceramic Society,
Building Materials Section,

20. Selander R.K. and Lemke G.A, 1984
"Dual Robotic Ceramic Handling System"
Society of Manufacturing Engineers,
Anaheim, California

21. Guha J.K, 1984
"Application of Robotics in inspection and manufacture of ceramic products"
Am. Ceram Soc Bull.
63 (9) 1179.

22. Graf B and Nicolaisen P, 1982
"Industrial Robots and their Operation possibilities in the Ceramic Industry"
Interceram.
31 (3) 209.

23. Rafferty J, 1982
"Automation and the Use of Robots in Kiln Car Handling"
Interceram
31 (3) 204

24. Anon, 1986
"Advanced Robot Systems sprays Vitreous Enamel for UK's Leading bath tub manufacturers".
Product Finishing,
39 (6) 14.

25. Shivarov N S et al, 1983
"Application of Industrial Robot for palletization of packages of wall tiles in the Ceramics Industry".
Institute of Engineering Cybermatics and Robotics
Sofia.

26. Clayton C.G.A, 1986
"The Use of CAD/CAM for the Design and Manufacture of Moulds for Ceramic Products"
Ceramic Forum Int /Ber DKG
63 (4/5) 216.

27. Dillon M, 1988
"Use of DUCT at Josiah Wedgwood."
User Presentation - DUCT User group,
Fourth Annual Conference,
Cambridge,

28. Anon, 1987
"Automation in the pottery"
Ceramic industries Journal
96 (1062) 28

29. Kay R, 1988
"Introduction to experimental design"
Course notes
"The effect of manufacturing processes on Design"
The University of Sheffield

30. Mather Hal, 1987
"Logistics in Manufacturing"
I.Mech.E.
London.

31. McLoughlin P, 1988
"Observations on the Use of Computers in the Ceramics Industries"
Br. Ceram Trans J. 87 (4). 113.

32. Lutcke H, 1985
"CAD in Refractory Construction"
T12 Fachberichte.
Vol. 109, No 8.

33. Anon, 1985
"Heavy Duty Handling Robot Performs in the Wet"
The Industrial Robot.

34. Burchell K, 1982
"Development of Control Systems in the Heavy Clay Industry"
Br. Ceram Rev 49-52, 43.

35. Chabinsky J J and Eves E E, 1985
"The Application of Micro-wave Energy in Drying Calcining and Firing of Ceramics".
American Ceramic Society 7th Annual Meeting
May 1985

TABLE 2.1 The Ceramics Industries

Classification	Sales Value (millions)	Number of Companies	Additional Classification
Tableware	441.0	75	
Sanitaryware	143.0	15	
			Whitewares
Wall & Floor Tiles	112.0	15	
Engineering Ceramics			Industrial
Electronic Porcelains	87.3	31	
Electronic Ceramics			
Bricks	470.3	30	
Clay tiles	16.3	n/a	Heavy Clays
Pipes	66.0	n/a	

Employees

Whitewares & Industrial	54,000
Heavy Clays	17,700
Total	71,700

Export Value

Whiteware & Industrial	275.0 millions
Heavy Clays	13.9 millions

Sales Value

Whitewares	696.0 millions
Industrial	87.3 millions
Total including services	1000.0 millions
Heavy Clay	640.8 millions

Chapter 13

Computer-aided control of welding systems

T. L. FitzGerald, A. R. Powers, J. S. Smith and J. Lucas

1. INTRODUCTION.

The work described in this paper was commissioned by a company dealing with the undersea welding of large diameter pipes for the oil industry, using a diving bell.

Subsea welding is a complex process, requiring the knowledge and expertise of specialised welders, who are in short supply. Errors cause expensive delays and can produce pollution. Also, there is a limit to the depth to which welders can dive Virr (1). To reduce the probability of errors, and to increase the depth at which welding can be performed, welding robots are being developed. These can be clamped onto the pipes and programmed with various welding parameters. However, these robots do not have any perception of the pipes to be welded, their orientation with respect to the robot and each other, and welding fit-up - the vertical and horizontal gaps between the plates. It is important to know the fit-up as the welding parameters which have to be employed will depend on them; for example the larger the gap the more filler material will be required.

A vision system has been designed to evaluate the two fit-up parameters required. It has the advantage that it is not susceptible to electrical noise, and the host computer can be sited remotely i.e. in the diving bell or on the surface vessel'.

2. HARDWARE.

A block diagram of the hardware configuration is shown in Fig.1. The laser diode casts a line of light onto the junction of the weld preparation; this line is detected by the camera. The camera is set at an angle of approximately 60 degrees, this enables the camera to detect an image showing both the vertical and horizontal gaps (Fig.2.). The image from the camera is passed to a video frame store which is plugged into the host computer. Real-time and processed images are displayed on a video monitor. Fig.3. shows the equipment on a test rig.

2.1 Camera.

A CCD camera was chosen for its robustness, and its small size. The size factor was particularly important as the equipment has to fit in specially designed pressure-resistant cans whose internal pressure is maintained at 1 atmosphere, which are mounted on the robot frame. The overall size of the camera including lens was 4.5 cm in diameter by 12.5 cm long, the focal length was 25mm, the resolution 604 x 588 pixels, and the access time 40mS.

2.2 Laser Diode.

The light source used was a 30mW 780nm laser diode, chosen for size and price considerations. The diode was mounted on a plastic head within a brass tube containing the lens for focusing the beam and the electronics to drive the laser. The size of the diode and lens was 1.5 cm diameter by 3 cm long. The focal length of the lens used was 25mm. The whole unit was placed in another pressure-resistant can for attachment to the robot.

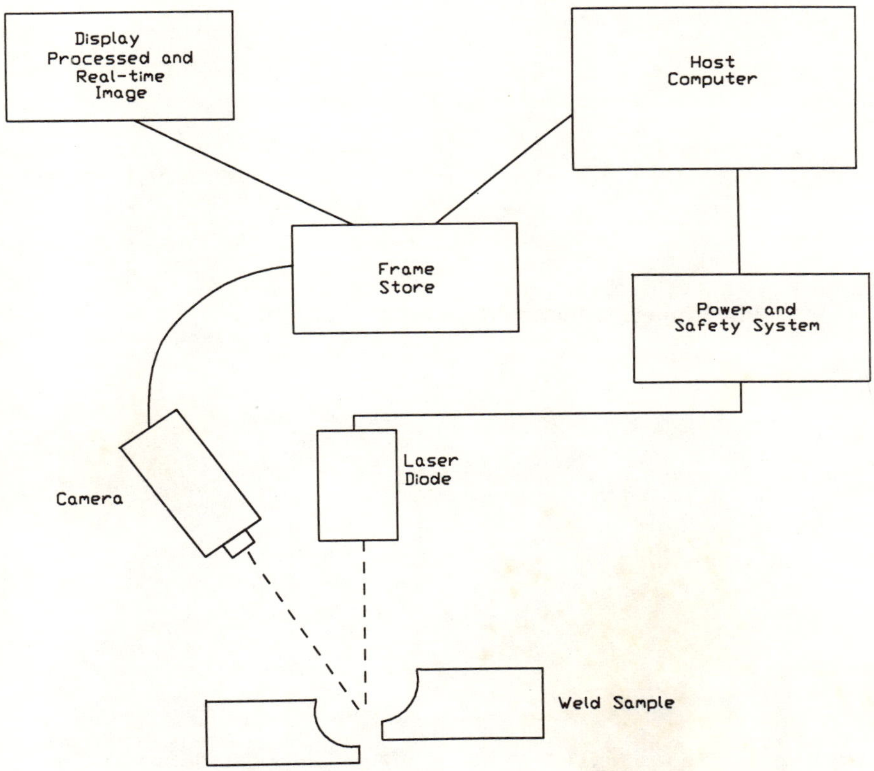

Fig.1 Hardware block diagram

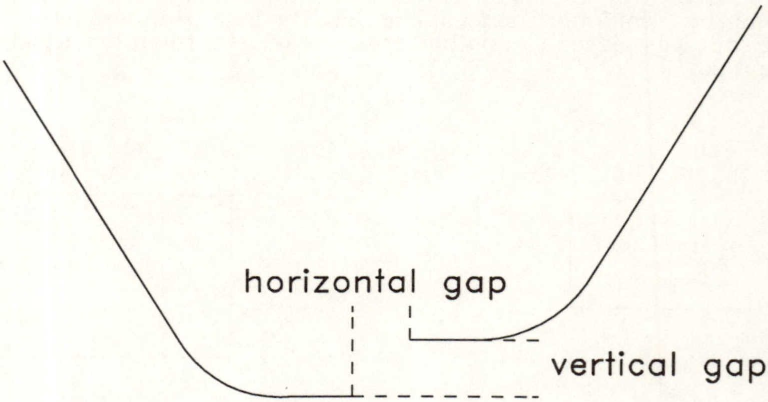

Fig.2 Image detected by camera

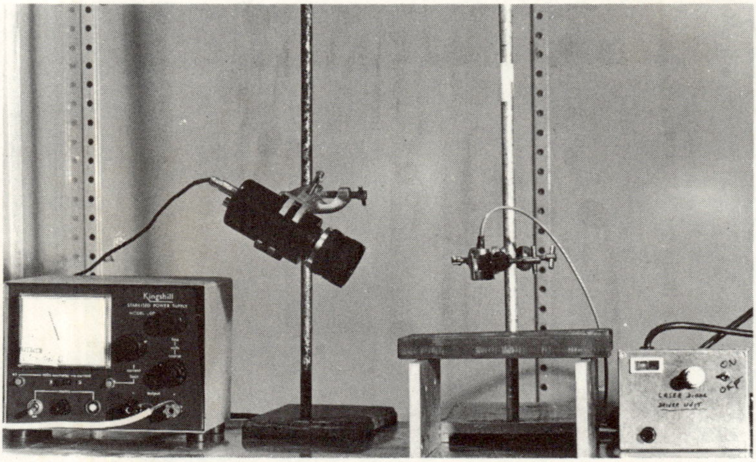

Fig.3 Equipment on test rig

2.3 Frame Store.

The frame store chosen was a Data Translation DT2853 board, which has a resolution of 512x512 pixels with 256 grey levels. The pixel access time is of the order of $^1/_{25}{}^{th}$ second. This board has sufficient resolution and speed to meet the requirements of the specification, whilst its cost was not excessive.

The disadvantage with this frame store is that it is memory mapped above the 1 Megabyte address range and therefore pixel access must be via a low-level BIOS call. This call can read a block of up to 32 kilowords of data from a given start address, but takes almost as long to read 1 word as to read 32 kilowords since most of its processing time is spent setting up registers for the move. Therefore when time is critical it is obviously better to read as many bytes together as possible and store them in a buffer. The frame store memory is arranged such that horizontally adjacent pixels are stored consecutively, therefore reading a horizontal line is much quicker than reading a vertical line since it requires only one BIOS call instead of 512. In order to make the algorithms more efficient it was therefore necessary to turn the image through 90 degrees so that horizontal line reading could be used (Fig.4.).

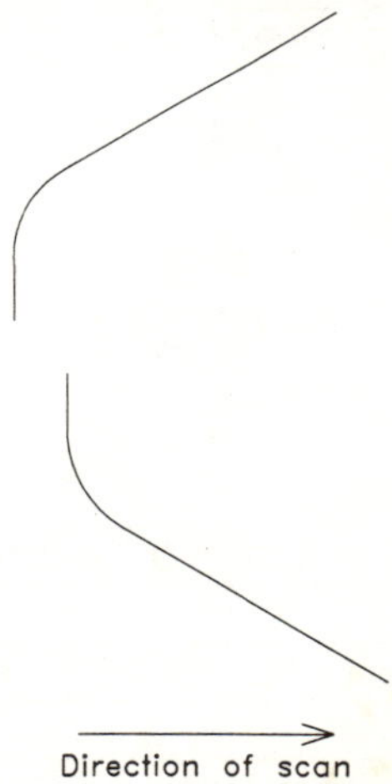

Fig.4 Image rotated through 90°

2.4 Host Computer.

The host computer used was an 80286 based personal computer with a 20 Megabyte Winchester disk drive and 640kilobytes of memory.

3. SOFTWARE.

All control software was written in the 'C' language with a small amount of assembly language for framestore access. A flow diagram of the software is shown in Fig.5.

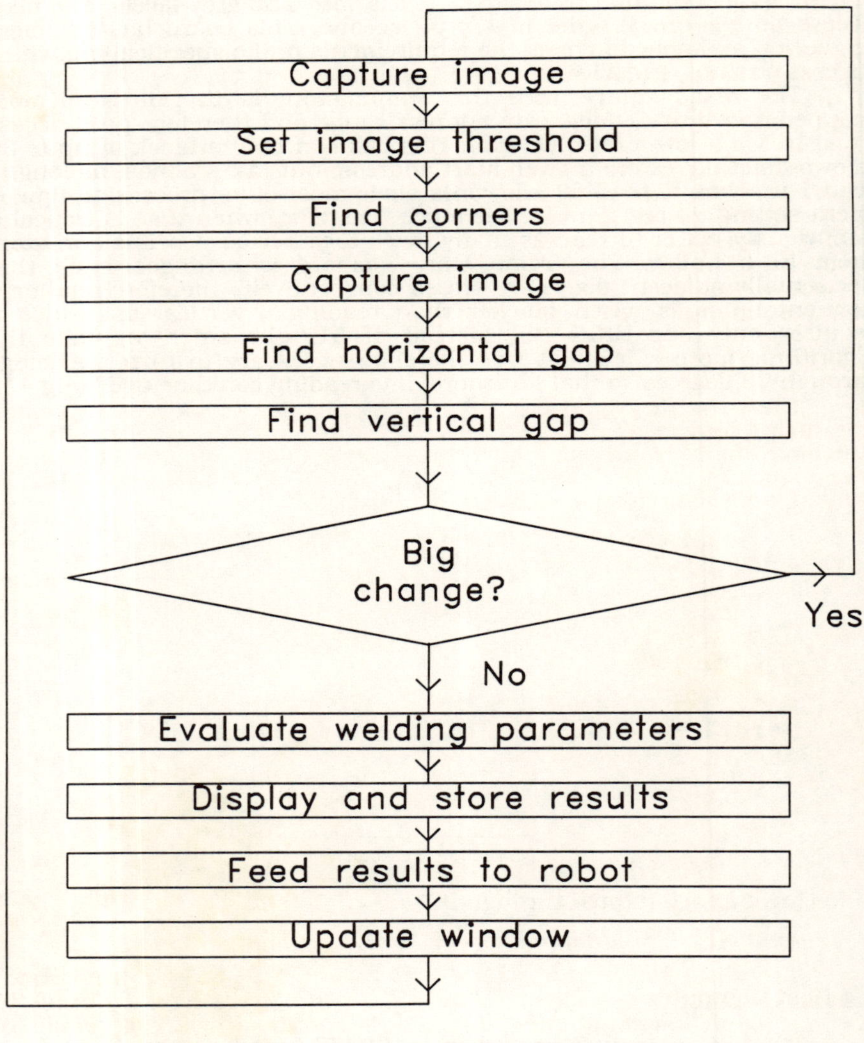

Fig.5 Software flow diagram

3.1 Framestore Routines.

The framestore could be accessed in two ways. Several 16-bit control registers are provided, these allow video input and output channels to be configured, hardware look-up tables to be set up, image capture, image display and cursor control. The memory can also be accessed via the BIOS routine described above and data read or written.

3.2 Thresholding.

The captured grey-level image is first converted into a binary image which is easier and quicker to process. A level must be decided upon, above which pixels are made white, and below which they are made black. To perform this action the computer reads the grey level of each pixel and then selects a grey level threshold such that 0.8% of the pixels are above this level. This value was found by experimentation to give the best results, due to the image being generated by a fine laser line.

3.3 Corner-finding.

Once the threshold level has been selected, work commences on identifying and measuring the vertical and horizontal gaps (the offsets). The next hurdle to overcome was the blurring of the outside edges of the image because of the limited depth of focus of the camera and also the limited focusing available with the laser diode (Fig.6.).

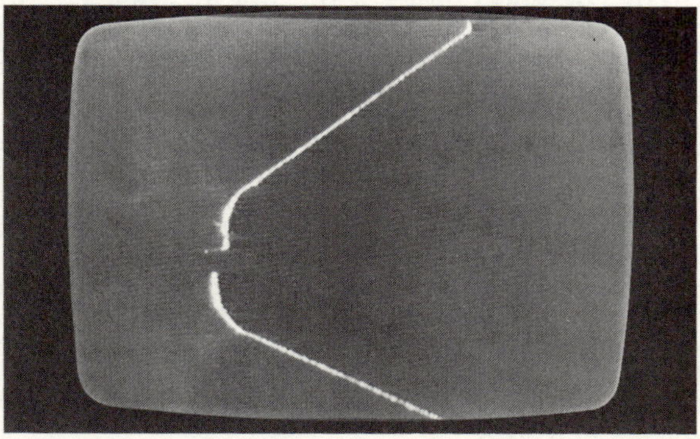

Fig.6 Image showing limited focusing

The area of interest is between the two corners of the J-preparation therefore anything outside these corners could be ignored. Thus if the corners could be found then the blurred part of the image could be deleted. A corner-finding algorithm was written which followed the laser line taking samples of the gradient across 40 pixels until a large change occurred, this was taken to be the corner. The algorithm was repeated for the other corner. The remaining algorithms could then be restricted to work within the two corners.

3.4 Finding Horizontal Gap.

The laser line between the corners is scanned, and the position of the middle white pixel of the line stored in an array. If a gap occurs a value of -1 is stored in the array for that position. When the scan is complete the information in the array can be used to find the sizes of the vertical and horizontal gaps. To find the horizontal gap, the array is scanned, at this point we are looking for a gap in the line i.e. the value stored in the array being -1. As breaks do occur in the line (usually < 5 pixels wide) due to dirty metal surfaces, we only consider a break in the line to be the gap we are looking for if it is more than 10 pixels wide. Once the gap region has been found, measurement of the gap is simply a case of counting the number of consecutive '-1's in the array. The other consideration is that of noise which shows up as white pixels, some of these may occur in what should be the gap region. By the same reasoning as before, we only assume white pixels in the gap region to be the start of the line if there are at least 10 consecutively. This checking means that the accuracy of the system is limited to measuring gaps of at least 10 pixels width, the pixel:mm ratio depending on the setup of the hardware, but being typically 25:1 i.e. the smallest measurable gap being 0.4mm.

3.5 Finding Vertical Gap.

Having found the size of the horizontal gap, the vertical gap must now be measured. The end points of the line nearest the gap are noted. The laser line is scanned backwards from one edge of the horizontal gap and the vertical position averaged over 10 points to eliminate noise. The same is done for the other edge and the difference between the heights of the two edges is the vertical gap.

If a horizontal gap is found, but there is no vertical gap found, the difference in the end positions of the lines is zero. If, however, no horizontal gap is found, there is still the possibility that a vertical gap exists, but the above method for calculating it will not work (i.e. there will be no end points found to start counting back from). Any discontinuity will most likely occur in the centre of the image since the image is almost symmetrical, thus a vertical discontinuity is looked for at the point in the centre between the corners. The last 10 points in the array approaching the centre from the top are averaged and then the same approaching from the bottom. The difference between these two values gives the vertical gap.

3.6 Windowing.

The routine described above is adopted for each scan; to speed up the process a 'windowing' technique is employed. All subsequent scans are limited to within 40 pixels of the limits of the gaps found in the previous scan. This way of updating the window allows the software to be operated in real time with a welding torch running around the weld. If the image wanders from the centre of the screen, the window will follow it. If, however, the picture makes a big jump between 2 scans the system will not catch the point of interest in its window and will re-initialise itself and scan the whole screen. The update rate depends on the size of the window because a major part of the time taken is to access the frame store. The update rate for a 40 pixel window is better than once every 3 seconds.

3.7 Calibration.

In order to produce dimensions in mm rather than pixels it is necessary to calibrate the system. This is done by running the software with a sample of known horizontal and vertical gap. A calculation is then performed to produce a mm:pixel ratio which can then be used in the calculations of the horizontal and vertical gaps. This was a preferred technique because it allowed changes of the angle of the camera to the workpiece and its lens system.

3.8 Welding Parameters.

The horizontal and vertical gap measurements are used to evaluate several welding parameters. This is done using a series of user-definable look-up tables, the default values being derived from data provided by the sponsoring company. As an example, Fig.7. shows the variation of welding current and wire feed rate with the gap width. Many of the welding parameters also depend on the angular orientation of the torch as it revolves around the pipe.

The start angle and the torch travel speed are input; from these and use of the real-time clock the angular position of the sensor can be evaluated. It is intended that this method will be replaced by an encoder.

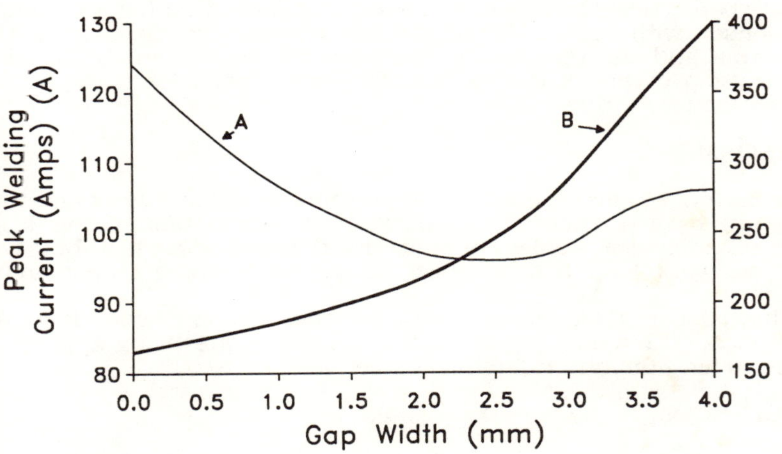

Fig.7 Graph of welding current and wire feed rate against gap width.

4. USER INTERFACE.

The system is menu-driven to allow operators not familiar with computers to operate it. The main menu (Fig.8.) allows the user to choose which of the functions listed above (calibration, defining welding context) is to be used. Other options available from this menu are that of sending the results obtained to a file, producing a hard copy of the results, or displaying a real-time image.

COMEX WELDING MONITOR v1.0
(c) A.R. Powers, J.S. Smith, & T.L. FitzGerald 1988.

1: Display real time picture
2: Calibrate
3: Define welding context
4: Start sensor without arc
5: Start sensor with arc
6: Send results to printer
7: Save results into file
8: Exit program

Fig.8 Main menu

Once an option has been selected from the main menu, the user is led through the necessary operations step by step. If either the option '4:run sensor with arc' or '5:run sensor without arc' is chosen, the system runs and the results are displayed on the screen. If the user wishes to stop or pause the system, the instructions for doing so are given as a menu.

5. ENVIRONMENT.

Experiments carried out using a pressure vessel at 30 bar pressure in a helium-oxygen atmosphere, showed no attenuation of the laser beam, no gas leakage problem in pressurised cans and no heat build up around the workpiece. A diver observed the trials using closed-circuit television.

The robot used was a four axis Cartesian machine which is servo-motor driven, TIG welding is the preferred technique being applied to 60cm diameter, 30mm thick stainless steel pipe.

6. CONCLUSION.

The system was tested and found to work successfully. It will be used as part of a fully automated welding system working below 600 ft depth, with no diver present. The automated system will have no decompression problems, and give lower running costs than manual systems.

7.REFERENCES.

1. Virr, L.E., 1987, Proc. IEE, 134, 547-576.

Chapter 14

Automatic laser-welding

N. F. F. Willmott, R. Hibberd, W. M. Steen and V. M. Weerasinghe

ABSTRACT

The laser can be a high quality welding tool. It has the further advantage in that it is one of the easiest welding systems from which to obtain in-process signals. This makes the laser an ideal partner for robotic welding applications. This paper reviews some of the in-process signals recently developed for laser-welding, in particular the use of an acoustic weld-monitor is described, as well as a recent developed non-visual seam-following system. The concept of an "intelligent" control-system using a real-time expert-system for control of the process is also discussed.

1.0 INTRODUCTION

As the field of laser-robotics continues to develop, the need for effective control systems that allow compatibility with advanced manufacturing environments increases. The area of laser-welding using keyhole welding finds more application as time progresses and hence control of this process provides an expanding field for research.

At the present time a computer integrated control system based upon a real-time expert system is being developed for a laser-robotic welding cell. A schematic diagram of the system is shown in fig.1.

The objective of the project is to produce automonous adaptive control of the process. Data is collected from the operator and various components of the system and fed to an expert-system. This uses the method of relational matrices, Li (1) to invoke rules and, according to the certainty of the input data, select appropriate instructions for the various controllers. The various components of the system, some of which have their own dedicated microprocessors, are able to communicate with the supervisory computer via the VME bus.

In-process signals collected include laser-power and mode structure, beam alignment with the transporting mirrors and nozzle, monitoring of the stability of the welding process, penetration of the laser beam into the workpiece and uniformity of the weld bead profile.

Two important parameters are the coupling of the laser beam to the workpiece and the positioning of the laser beam on the seam. The function, operation and characteristics of two devices developed to achieve this are now described.

2.0 KEYHOLE/PLASMA MONITORING

When the focused laser beam strikes the surface of the workpiece, reflection and absorbtion take place. If sufficient energy is absorbed the surface material will vapourise and ionise to form a plasma. This absorbs the laser beam energy as well and consequently becomes extremely hot, melting a channel in the workpiece. Some of the laser energy is reflected back up through the optics towards the laser cavity.

The power distribution across laser beam varies both spatialy and temporaly. When this strikes the mirror, localised heating occurs generating stress patterns which in turn generate acoustic signals throughout the mirror.

The signal is dependent on the mode and power of the laser beam. It has been shown by Weerasinghe (2) that a signal is generated by the back reflected energy from the workpiece and that when a keyhole is formed the signal falls appreciably. Since a low signal is also obtained when the defocused beam impinges on the workpiece the acoustic signal on its own is insufficient to determine the presence of the keyhole. Thus, an optical sensor is employed to observe the plasma generated when a kehole is formed.

2.1 Apparatus

A schematic diagram of the system is shown in fig.2. The system uses two sensors, an ultrasonic (1MHz) broadband acoustic-emission sensor and a optoelectric sensor. The acoustic sensor is mounted on the rear face of the final mirror before the lens, the optical sensor near to the nozzle. The optical sensor, which is sensitive to the visible spectrum, is screened by a blue optical filter. The acoustic signal is amplified 80 dB and converted to an RMS signal. The signals are input to analogue to digitial converters which input the digital signals to a 6502 microprocessor.

The microprocessor records and analyses the data. By setting suitable thresholds for each of the signals they are converted to binary signals. Using Boolean algebra the presence of a keyhole may be deduced.

2.2 Experiment

The experiments performed consisted of scanning the workpiece with the laser beam whilst changing the position of the focus. The focus of the laser beam was brought down to and through the surface of the workpiece. The focus starts at 6 mm above the workpiece and descends to 3.5 mm below it. The laser beam traverse speed was 12.5 mm/s, the power level 1.3 KW. The workpiece was 2mm mild steel. This is illustrated in fig.3.

This resulted in the energy intensity at the workpiece increasing, reaching a maximum and then decreasing. Provided that sufficient power is available, a keyhole will be created. Whilst this is taking place signals are collected by the two sensors.

2.3 Results and Discussion

The signal shown in fig. 4 shows the variation in acoustic signal. The results show that the signal rises sharply as the focus nears the surface of the workpiece until a keyhole is generated at which point the signal falls sharply. Whilst the keyhole is maintained the signal level remains low. When the keyhole fails the signal rises sharply again.

As can be seen, the signal oscillates as the focus descends to the workpiece. The oscillation in the signal and penetration is caused by the plasma rising and blocking the beam, thus temporarily extinguishing the keyhole, increasing the back-reflection. The signal then remains fairly level when a stable keyhole is maintained. Over this region the focus position varies by 2mm. As the focus continues to descend the power density falls and the keyhole is extinguished.

The results of an experiment when the acoustic and optical signal are recorded are shown in fig.5. Readings were taken every 40ms and hence some detail is omitted. The signal is in fact very noisy, due to the mixture of randomly varying frquency componets. Faster signal recording and time-averaging will improve the signal to noise ratio. It should be noted that the experiment produced continually changing process conditions. This would be expected to affect the signals as well.

Results of previous experiments have shown that the ratio of the acoustic signal either side of the keyhole region , to that within it, is of the order of 7.2 : 1 with a standard deviation of 1.1. Setting a threshold level allows the signals to be converted to a binary form. Examining the condition when the acoustic signal is low but the optical is high, allows the determination of the keyhole region. This is also illustrated in fig. 5. The reason for using this method is that other factors other than keyholing, such as loss of laser power of seam-tracking error, may result in a drop in the acoustic signal level.

Thus once the system has been calibrated it is possible to determine the range of focus position that allows keyholing to take place, and a calculation of the optimum position of focus, automatically. If, as part of an initialisation sequence, the focus is scanned, this will allow self-learning by the system.

In conclusion it is shown that the system can be used to deduce the presence, or failure, of the keyhole, and that it offers the potential, after further development, for use in height-sensing and self-learning for optimal positioning of the beam-focus.

At the present time more detailed analysis of the acoustic signal is being carried out which is expected to lead to a means of determining the status of other process and quality parameters. This in turn will allow improved control of the welding process in a laser-robotic system.

3.0 SEAM-TRACKING SYSTEM

The seam tracking system locates the seam ahead of the laser beam in order to allow the beam to be positioned upon it and welding performed. The system is illustrated in fig.6.

The system uses a pneumatic sensor to provide a non-contact means of detecting lap and butt seams. The pneumatic sensor has the advantage over other sensors that it is not affected by the material properties or the surface characteristics of the workpiece and is sensitive to very narrow gaps, such as those found in butt-seams for laser-welding.

The sensor issues a gas jet whilst scanning the workpiece. The discontinuity in the surface created by the seam produces a pressure pulse in the gas flow, which is detected by a pressure sensor. The form of the signal obtained s the sensor scanned over a lap-seam of 2mm mild steel is shown in fig.7. The signal is processed electronically to convert it to a digital pulse.

The microprocessor controls the scanning of the sensor via a stepper motor controller and the position of the CNC table via another controller, to produce a 3-axis system. Using the known location of the sensor when the seam is detected, it's position may be defined. The seam can then be mapped out as the sensor progresses along it.

The laser beam is at the center of rotation of the sensor. The microprocessor positions this point on the seam allowing processing to take place. At the same time the position of the seam ahead of the laser is determined by the scanning sensor.

Height sensing can be achieved using a static pneumatic sensor following one edge of the seam. This too would be positioned by the microprocessor via a controller. Various configurations of pneumatic sensor are available for height sensing (3).

4.0 CONCLUSION

The keyhole/plasma monitoring, seam-tracking and overall system described will allow greater automation and hence, enhance the performance of the lase-robotic welding cell. The use of a real-time expert-system will allow the incorporation of 'rules-of-thumb' used by the operators in the decision-making process and permit self-learning by the system. These together will assist in the optimisation of the laser-robotic welding process.

5.0 REFERENCES

(1) Real-time expert system for supervisory control of laser cladding, Li Lin, W.M. Steen, R.H. Hibberd, V.M. Weerasinghe, Proc. Conf. ICALEO 87.

(2) Weerasinghe, V. M. and Steen W. M. , In-Process Monitoring of Laser Processes, Proc. Conf. ICALEO 85.

(3) Schweisstechnik, Berlin 34 (1984), 7, 299-301

6.0 FIGURES

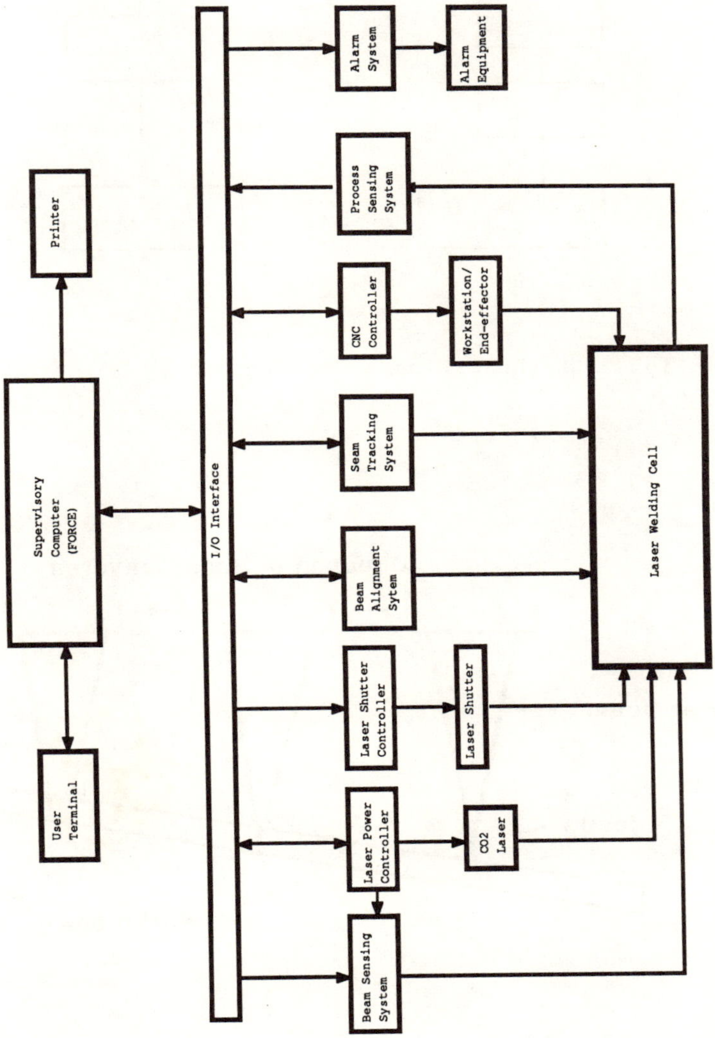

Fig. 1. System Hardware

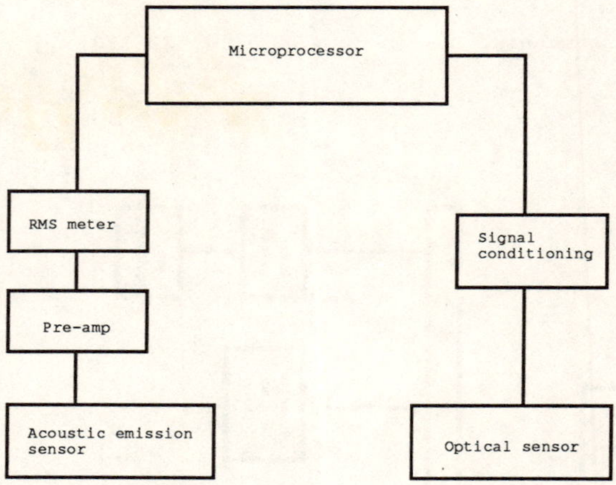

Fig. 2. Keyhole/Plasma Sensing System

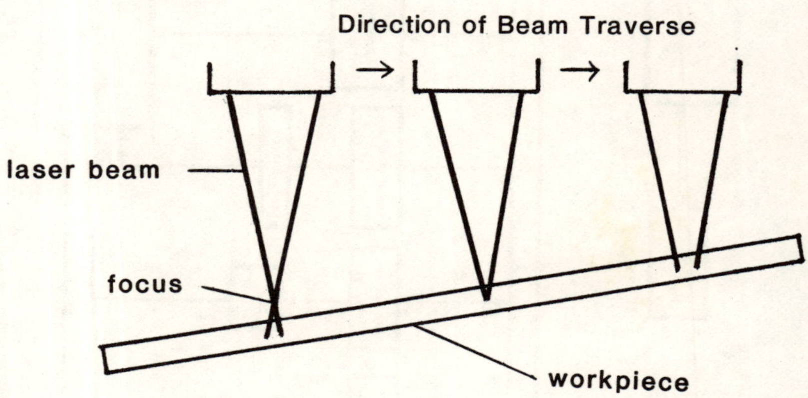

Fig. 3. Experiment

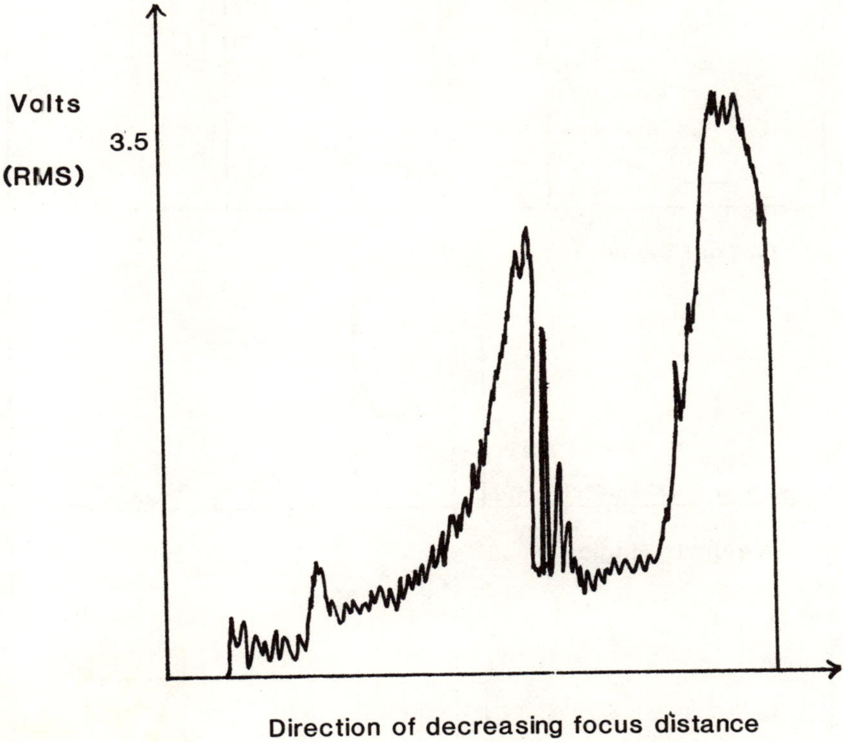

Fig. 4. Acoustic signal

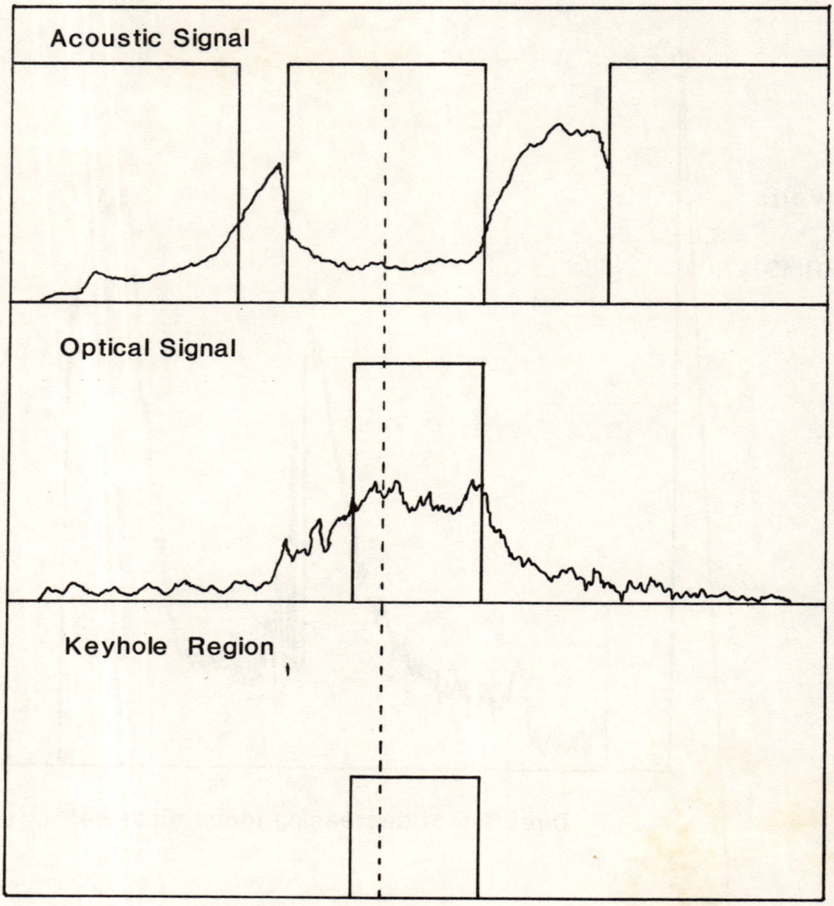

Fig. 5. Processing of Acoustic and Optical Signals

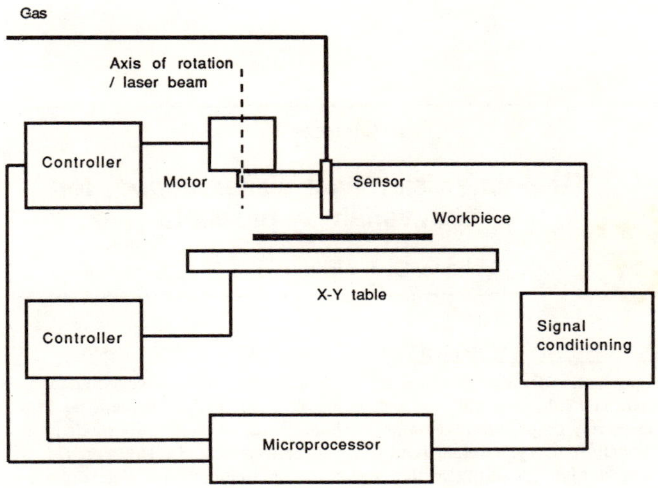

Fig. 6. Seam-Tracking System

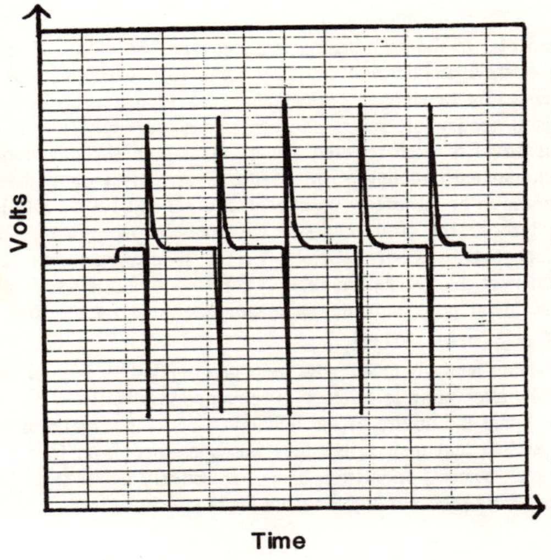

Fig. 7. Signal from Pneumatic Sensor

Chapter 15

Real-time vision analysis techniques for high variability products

R. Gibbons, D. J. Williams and D. A. Green

1. INTRODUCTION AND BACKGROUND

In food production the identification and orientation of objects by visual methods is an important tool for automation of the assembly process. However most of the conventional techniques used for object recognition rely on products which have a high degree of conformity of shape as well as 'clean' images. Additionally the objects often need to be accurately located. This is not the case in the food industry as there is high variability in the product and placement on a plate or in a container needs only low accuracy. However the speed of processing is paramount as rates of several hundred items per minute are not uncommon. An investigation was carried out to compare various vision analysis techniques, all operating upon the same task. The relative processing demands, and thus the time taken for each technique could be found as well as a measure of accuracy of the analysis. Earlier work on food products by Davis (1), has concentrated on such techniques as Hough Transforms. These can give fast reasonably accurate results but are only really suitable for circular or uniform products.

2. THE TASK

A problem particular to food processing is the natural variability of raw material. An operator will assess the 'shape' of a product and mentally compare it with standards built up by experience before deciding what action to take, i.e. orient and pack or reject. A machine vision system must emulate these actions at comparable speed. To compare shape analysis techniques a ''real'' task was needed. Coated meat products were selected as typical of both shape variation and of products that lend themselves to automated handling. The product is roughly kidney shaped without any definite corner features but needs to be accurately orientated in order to be filled into its pack without overlapping the edges. Fig.1 shows the product model and examples of variation found in practice. Due to the variation in product shape it is only possible to define their orientation to an accuracy of 10° - 20°. The speed of production of this product is about 3 per second and although the mechanical placing is unlikely to be at this rate it gives an order of magnitude value for the time available for the vision task, i.e. in the order of 0.1 - 1.0 seconds.

Four types of vision analysis techniques were investigated as to their suitability to perform this task. This was done by attempting to orientate a reference product from 36 known positions with 10° increments. To investigate the ability of the methods to orientate real, variable product ten samples were taken of true product and their orientation was attempted from four different positions of 90° increments. (The initial 'correct' orientation being subjectively assessed by 'eye'.) As the size of the product varies, all the techniques investigated were made size invariant.

3. EQUIPMENT

The equipment used for this analysis comprised of a Vision Dynamics VCS 512 frame store and processor. This is a VME backplaned device which was hosted in a Electronic Modular Systems 68020 system, upon which the programs were developed. The language used was Microware C with Vision Dynamics VCS standard libraries. The objects under inspection were placed on a rotary table under a Panasonic colour camera, or pre-recorded video tape sample images were used. For the noise sensitivity analysis tape of a typical production line was used. The size of the images was fixed so that a window of 212^2 contained the image in all orientations. This reduced the overall processing time but kept the relative times of the methods constant.

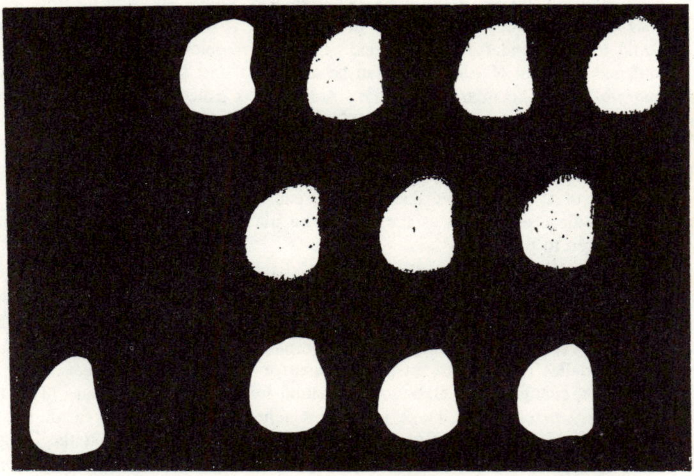

Fig.1 The product model and examples of product variation

4. RESULT FORMAT

For each analysis the following results are given; the time taken, the value of a parameter which is altered to affect the time the analysis takes, a measure of the accuracy which is based upon the sum of the mistakes made divided by the total number of angles measured. For the first two analyses there is also a confidence factor, which gives a measure of how significant the matching is.

5. DUBOIS & GLANZ ANALYSIS

This work is based upon Dubois & Glanz (2) and Kashyap & Chellappa (3). Here the boundary is approximated by an ordered sequence of the lengths of N angularly equispaced radius vectors projected from the boundary centroid to the boundary as shown in fig.2. The work followed this approach but used the object centroid instead of the boundary centroid as this does not require the object's boundary to be identified.

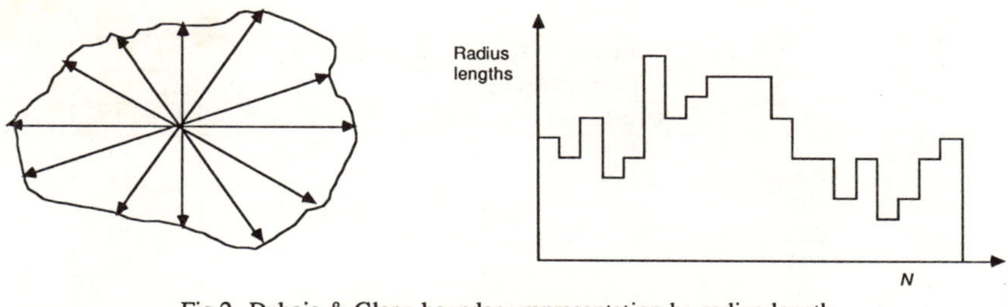

Fig.2 Dubois & Glanz boundary representation by radius length

As few food products have high radius of curvature concave boundaries to the extent of causing re-entrant problems as those dealt with by (2) this modification to the method was ignored to keep the speed of operation to a maximum.

To implement the method the following procedure was taken, this is significantly different to the boundary following method employed by (2) as it does not require Freeman chaining. Boundary identification is both time consuming and liable to error.

From a thresholded image, the object centroid was found, as well as the maximum and minimum values of the x and y co-ordinates of the object. From these a maximum radius length (from the centroid to the boundary) can be found. Using a pre-calculated look-up table supplying the function of 'sin' and 'cos' for N (this value can be altered) a start and end point for a vector can be calculated along which two boundary points are. A routine is called to plot along this vector and to find the points from which two radius lengths, 180° apart, are calculated. The object's radius vectors are found by repeating this process $N/2$ times and incrementing the angle of the vector by $360/N$.

To find the orientation of the object this set of radius vectors is matched N times with a reference set with offsets of 0 to N. The absolute differences for each matching are summed and the offset (ω) for the smallest sum taken as the orientation match. The orientation of the object to the reference object is taken to be $360\omega/N$.

To achieve the 10° precision specified N must be set to 36 or more. A major factor in reducing the time taken for analysis is to keep N as low as possible. This is because N^2 comparisons must be performed during the matching. Thus, for the experiment, N was set to 36, i.e. 36 radius lines were used. (As the test object is a 'rotund' object there would be little gain in increasing N, as the ΔN amounts would be small. This would not be the case for more angular products.) To gain a time against accuracy result, changes were made to the method for finding the centroid. (This takes a large time overhead as every pixel in the image must be scanned.) A 'scan line' factor was introduced, which effectively reduces the definition of the image. Instead of inspecting all the images pixels to obtain the centroid, only one line in λ would be inspected and only one pixel in λ along that line. Thus, the number of pixels needed to be accessed would be reduced by λ^2, reducing the processing time by a similar factor. However this has the effect of skewing the centroid from its true position proportional to λ^2 which similarly affects the radius vector readings.

5.1 Results

The results of the analysis are shown in table 1. The confidence factor shown is calculated by the following formula:

$$Confidence = 100\left(1 - \frac{N Score_{\min}}{\sum_{i=1}^{N} Score_i}\right)$$

and is taken to be a measure of certainty of the match. Fig.3 shows a typical matching of the object. Fig.4 shows the results from the product model in the form of a graph plotting times, accuracies and errors scan-line separation. The times shown are only representative times as they can vary from image to image as the values for matching change. The results from the true product are shown in fig.5.

Table 1 - Dubois & Glanz Results ($N = 36$)

λ	Time (msec)	**Product Model**		**True Product**	
		Confidence (%)	Error (%)	Confidence (%)	Error (%)
2	840	82.44	0	73.32	0.93
5	580	82.75	0	73.32	1.08
10	540	80.22	0	68.18	1.08
15	530	75.89	0.15	57.23	1.23
20	500	71.78	0.31	41.15	18.6
25	490	63.17	1.39	-	-

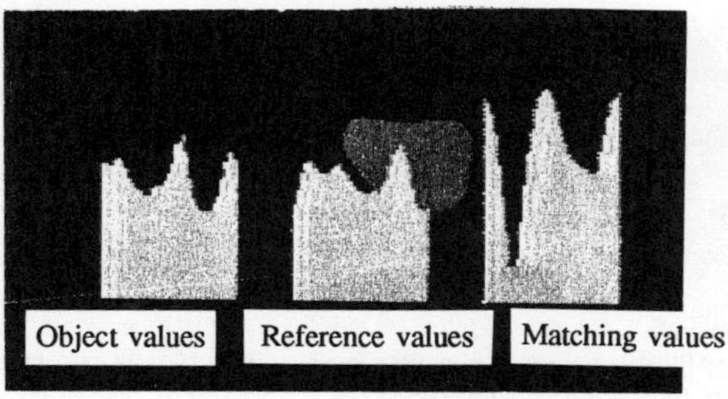

Fig.3 Matching an object, Dubois & Glanz

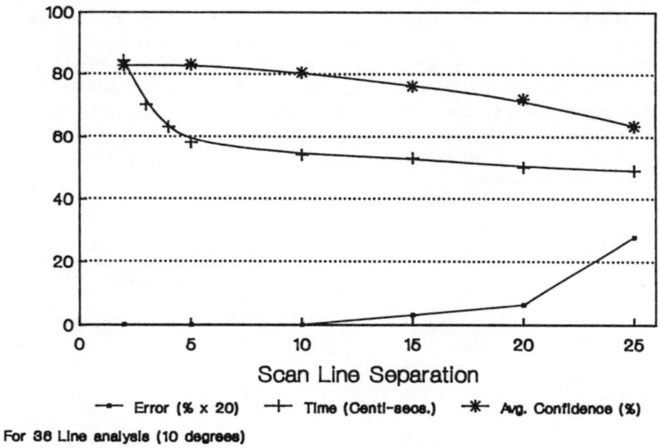

Fig.4 Dubois & Glanz, results

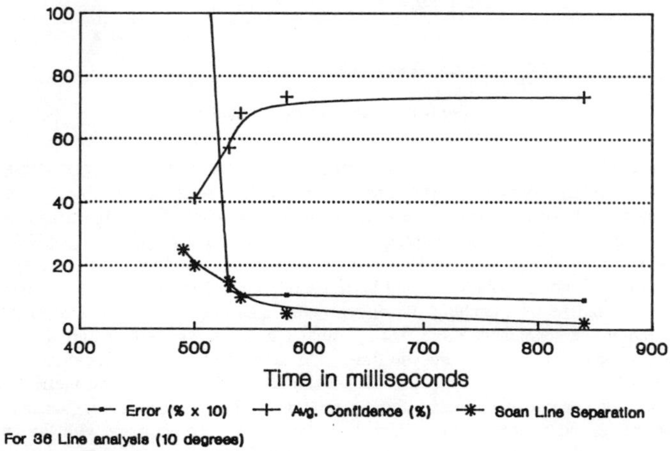

Fig.5 Dubois & Glanz, variability results

5.2 Comments

This method of analysis shows a rapid deterioration of accuracy as λ, the scan line separation, increases. For the size of sample shown this occurs about λ=20. This is a result of the determination of location of the centroid becoming inaccurate so that the plotted radius vectors' lengths are adversely affected. However, at this value of λ not much reduction in time for analysis is gained from lowering it further. This is because most of the remaining processing time is spent on the matching which is fixed for the 10° accuracy required. The variability results echo this.

The match score was calculated as the sum of the absolute differences between the reference object and the object under test. This saved time over a more exacting method of taking the RMS (root mean square) values. An investigation of taking these values was found to be no more accurate, and on a particularly noisy picture could even be less accurate, as one large erroneous radius length value would be amplified by a RMS approach.

6. JUVIN & DUPEYRAT ANALYSIS

This analysis uses a contour encoding technique. It is based upon the work of Juvin & Dupeyrat (4). The technique is to Freeman chain code the object (5). This tracks the edge of the object and assigns a value to each pixel on the perimeter dependent on the direction of this pixel from its previous adjacent perimeter pixel. Once the Freeman chain has been calculated it is split up into Λ equal lengths. (The odd pixels being evenly distributed.) The angle between the adjacent lengths of perimeter ($\delta\theta$) is then calculated. Once the object has been circumnavigated in this way a set of $\{\delta\theta\}$ is produced. (The sum of which is 2π.) This can then be matched Λ times with a reference set of $\{\delta\theta\}$ in a cumulative way with offsets from 0 to Λ in the $\{\delta\theta\}$ reference set as starting points and a best match found as shown in fig.6.

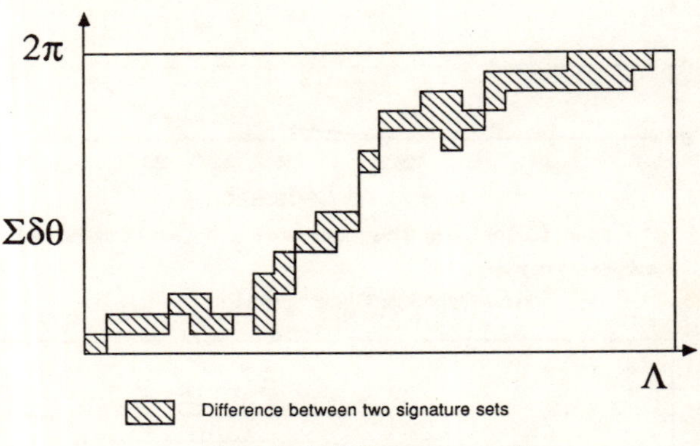

Fig.6 Juvin & Dupeyrat, matching

This method was found not to work satisfactorily as the method for generating the Freeman chain code used always started the encoding at the top left pixel of the object. As the object is rotated this start point does not progress evenly along the perimeter of the object, and in fact can miss out concave parts of the perimeter completely, hopping from point to point. A possible way to counteract this would be to always start the Freeman chain encoding from a point fixed relative to the centroid of the object. Another solution would be to encode the reference object in as many orientations as required, (R), and to try matching the object under inspection's $\delta\theta$ values to each reference set once. This requires considerably more 'set-up' time but, as it does not require the centroid of the object to be calculated saves during the run time. The amount of matching required is a factor of $R\Lambda$ as opposed to Λ^2 for the original method. The maximum accuracy of the first method will be dependent on Λ (but not linearly incremented about a rotation of 2π) whereas for the second the maximum accuracy for the second method is linear and dependent on R. To compare with the Dubois method, and to maintain the 10° precision required, R was fixed to 36. Λ was varied to see the effect of CPU time on accuracy.

6.1 Results

The results of the analysis for varying Λ are shown in table 2. The confidence factor is calculated as for the Dubois & Glanz analysis but substituting R for N. Fig.7 shows a typical matching. Fig.8 shows the results in the form of a graph as before. Again the figures given for time are somewhat dependent on the actual values of $\delta\theta$ being compared. No experiments were conducted for tolerance for the variability of the product due to the poor response of the method to the test object.

Table 2 - Juvin & Dupeyrat Results (R = 36)

Λ	Confidence (%)	Error (%)	Time (msec)
12	85.8	3.55	265
18	72.4	3.70	291
20	84.6	1.85	300
24	76.0	1.70	324
28	80.8	2.78	345
30	74.0	2.47	364
32	70.4	1.85	370
36	66.4	2.78	379
40	66.2	3.70	403

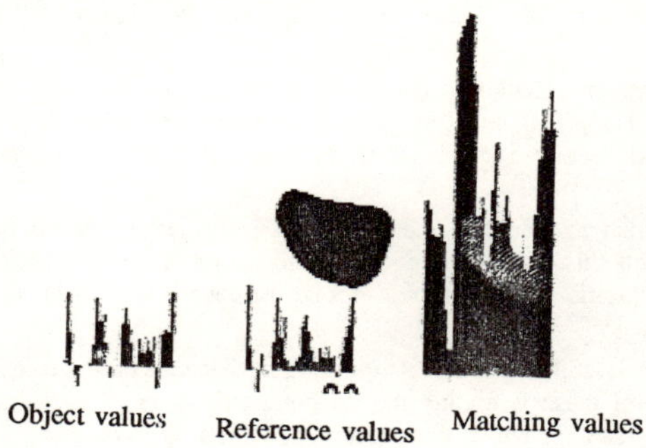

Fig.7 Matching an object, Juvin & Dupeyrat

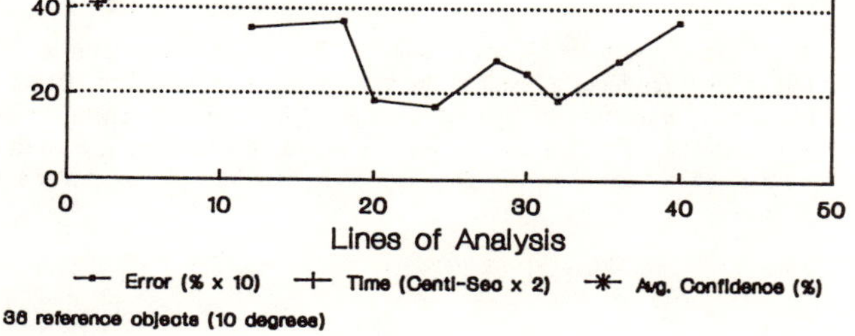

Fig.8 Juvin & Dupeyrat results

6.2 Comments

This method proved to operate fairly quickly (~300ms) but unfortunately is too prone to noise to operate reliably. (By noise it is meant distortions of the image, often due to non-uniform lighting, or small objects in the background of the image.) The perimeter of the object needs to be identified rather than the area as in other forms of identification. This saves time as the amount of pixels being accessed is reduced but lowers the amount of information being used to identify the object. While this reduction is probably fine for clean images, or objects with a high degree of conformity this method fails to maintain accuracy when variable, noisy objects are attempted to be analysed. Again an investigation as to the improvement of results by using the RMS difference between the $\{\delta\theta\}$ sets. A marginal improvement was found but it was not significant, nor did it justify the consequent increase in analysis time required.

7. BULK ANALYSIS

Whereas the first two methods have concentrated upon using parameters to describe the object's perimeter in some form this method relies upon a simple measurement of bulk characteristics.

During the early sixties vision analysis was usually performed on coarse, low pixel count images, Hu, (6). This was partly due to the limitations of both camera and processing resources and the end results were consequently imprecise. For the food industry there is a similar set of requirements, processing resources are limited by time and the precision of the result is lower than usually required by vision inspection tasks. This means that some techniques used then are suitable for this orientation task.

(6) bases his work on calculating central moments of an object of varying powers. This was further investigated by Luo, Tsai and Lin, (7) as a rotational invariant object recognition system, by use of a object 'visual feature vector' which is a set of five parameters generated from central moments.

The processing load to calculate the higher orders of these central moments is large and (7) have attempted to lessen this by only conducting the operation on the perimeter of an object. This will work for clean, isometric objects but will not be the case if size varies or if the perimeter is much altered by noise.

Despite this there is still some useful information that can be gleaned from the lower orders of central moments. One of these is finding the principal axis of an object, that is a line going through the centroid with slope $\tan\theta$ for which the moment of inertia, is minimum. This has two solutions for θ which are 180° apart. This conveniently gives a measure of orientation, although θ has two values. To try to rationalise this the distance from the centroid to the perimeter in both directions is calculated. The test object has a fairly distinct difference between these two values, being 'pear-shaped' in form, and thus gravitating the centroid to the 'fat' end. With the aid of a reference object an orientation of the initial principal axis offset can be found so relative rotations can be calculated.

To investigate the effect of decreasing the time required for calculation a similar scan-line factor, λ, to that used for the Dubois & Glanz analysis was used both for the location of the centroid and central moments calculations. The time taken for the analysis is proportional to $1/\lambda^2$ so varying λ has a great effect on speed.

7.1 Results

The results of the analysis are shown in table 3. The results are presented with two values of accuracy. As this method produces a result to the nearest ±1°, as opposed to 10° for the previous two analyses, it is not a true comparison to measure the sum of these errors. Therefore all the errors have been rounded to the nearest 10°. This usually gives a slightly higher accuracy result but a more comparable reading. Both values are shown in graphical representations of the results and variability sensitivity in figs. 9 & 10.

Table 3 - Bulk Analysis Results					
λ	Time (msec)	Product Model		True Product	
		Error (%)	Equiv. Error (%)	Error (%)	Equiv. Error (%)
1	1000	1.28	0.46	1.53	0.54
2	470	1.33	0.15	1.65	1.23
4	260	1.42	0.77	1.62	0.54
8	170	2.54	2.03	2.70	1.92
16	120	3.95	3.78	6.89	7.17

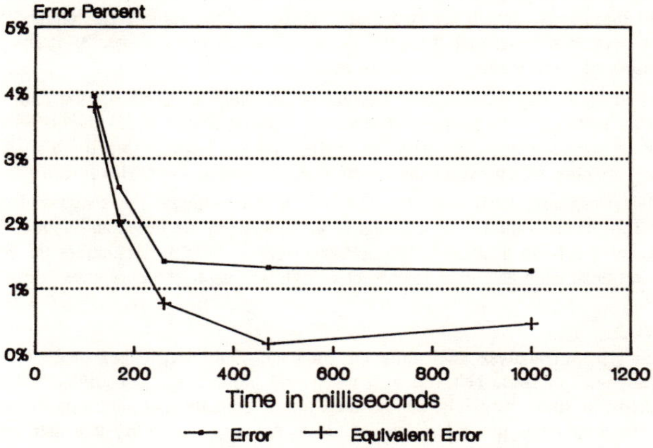

Fig.9 Bulk analysis, results

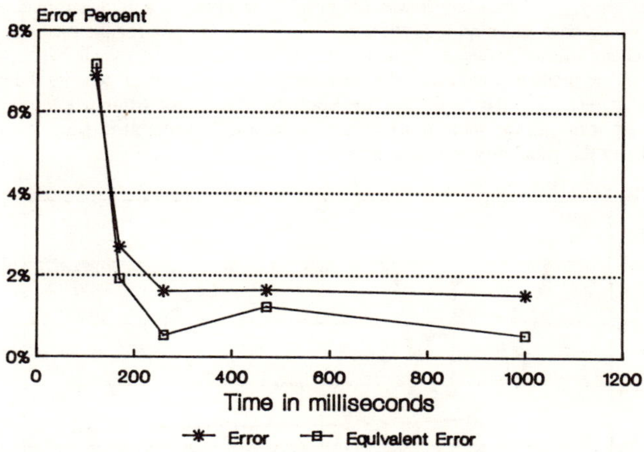

Fig.10 Bulk analysis, variability results

7.2 Comments

The results for the bulk analysis are accurate to about 1.5° for orientating the test object over a wide range of analysis speed until about 250ms when they deteriorate rapidly. At this point only 12.5% of the object is being used to generate the moment invariants so it is not unexpected that the accuracy is low. The 'cut-off' point is similar for the variability tolerance. There is some background inaccuracy present in this method. This indicates that the principal axis may be easily displaced from its usual position by changes to the product shape. This is more likely the more circular the product is. Another problem arose with finding the correct direction of orientation, i.e. whether θ_1 or θ_2 should be taken. At lower resolutions this proved a major problem and the accuracy suffered. Again this is very dependent on object shape.

8. PRIMAL SKETCH

The curvature primal sketch implementation is based upon the work of Asada & Brady (8) and is a form of recognising objects by means of significant changes in curvature of their perimeters. (8) show that generic shapes have generic features identified as 'corners' and 'smooth-joins'. These are further combined into 'ends', which are two nearby corners of the same sign, 'cranks' which are two nearby corners of opposite signs and 'bumps' or 'dents' which are two nearby crank changes. The ordered combination of these features is used for object recognition.

(8) conducted their analysis with hard angular objects such as a Warrington hammer, which has well defined corner features. The analysis depends on filtering $\delta^2\theta$ to detect the corners. These readings are very noise sensitive being a second derivative value. Furthermore the test object, has only small changes in curvature which makes the isolation of the primal sketch features difficult.

Trial analysis was conducted upon the test object to investigate the shape and signal to noise ratio of $\delta\theta$ and $\delta^2\theta$. These values were calculated working from the Freeman chain code in a similar way to that used for the Juvin analysis. Despite using various filtering techniques the $\delta^2\theta$ plot proved highly variable and unsuitable as a base for object recognition using the curvature primal sketch.

9. CONCLUSIONS

Of the four types of analysis used only the Bulk and Dubois & Glanz methods produced good repeatable results. Both of these behaved well when real product was substituted for the test object. Both these methods in some way rely on the area rather than the perimeter of the object and this reflects that an area reading is much less susceptible to noise or variability than perimeter. However the problem with an area method is that is will potentially require more processing time as more pixels are accessed. However the Dubois & Glanz approach quickly reduces the area measurements to a one-dimensional array. Unfortunately it was not possible to obtain a confidence reading for the bulk analysis. This would have given an indication to the tolerance to noise for the method. The results of the Dubois & Glanz method, when in error usually missed the correct match by one offset, or 10°. The variability of the product could cause the bulk analysis, when in error, to miss the correct orientation by a large amount. For the 'real' task the constant small error exhibited by the bulk analysis would not be significant but the sudden swings are hazardous. The combined accuracy/time results are shown in fig.11 and the variability results in fig.12.

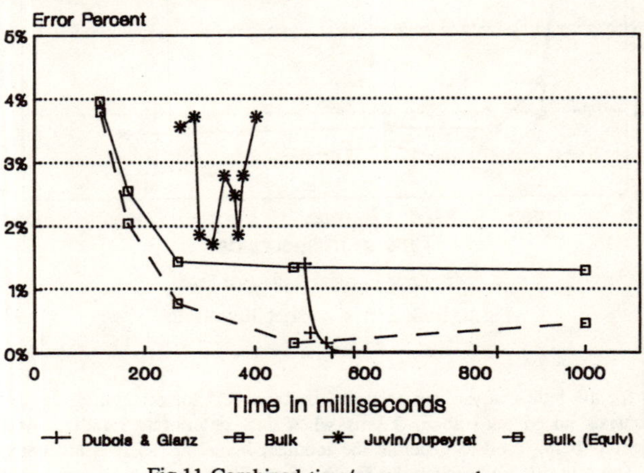

Fig.11 Combined time/accuracy results

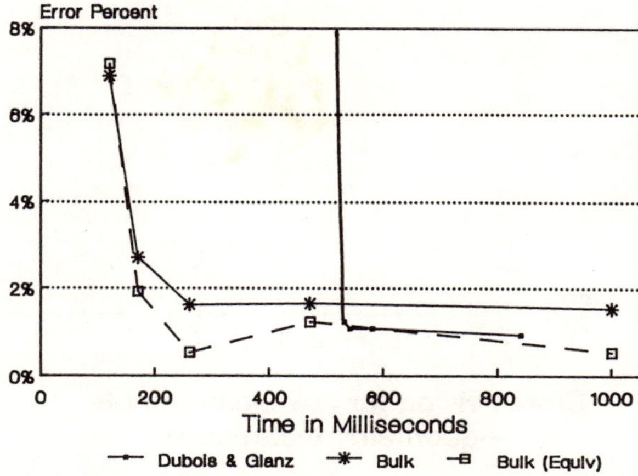

Fig.12 Combined variability tolerance result

One way of reducing the error is to combine two or more methods of orientation by use of rules. If two or three measures of orientation were taken and the results conflict then rationalisation of results is required. To resolve this conflict a rule-set must be implemented which would make an absolute decision as to the orientation. An approach would be to look at the scores of successive orientation matching and to see if the match is an isolated result - in which case it can be discarded, or if it was part of a near-match peak, in which case the result would be valid. Using rule-based techniques for the high-level interpretation of results of vision can also help discard wild answers, such as false object detection, or noise falsified readings. This is because knowledge of the product , such as the expected size and shape variance, is built into the analysis just as a human operator or quality controller has an understanding of the product being inspected.

10. ACKNOWLEDGMENT

The authors would like to thank Unilever UK Central Resources for supporting this work.

REFERENCES

1. E R Davis, 1985, 'Radial histograms as an aid in the inspection of circular objects', IEE Proceedings, 132, Pt. D, no 4, 158 - 163.
2. S R Dubois and F H Glanz, 1986, 'An Autoregressive Model Approach to Two-Dimensional Shape Classification', IEEE Transactions on pattern analysis and machine intelligence, PAMI-8, no 1, 55 - 66.
3. R L Kashyap and R Chellappa, 1981, 'Stochastic models for closed boundary analysis: Representation and reconstruction', IEEE Transactions on information theory, IT-27, 627 - 637.
4. D Juvin and B Dupeyrat, 1984, 'An Heuristic Method of Classification and Automatic Inspection of Parts: The Recognition System of ANIMA', Proc. 4th Conference on Robot Vision, 361 - 370.
5. K S Fu, R C Gonzalez & C S G Lee, 1987, 'Robotics - Control, Sensing, Vision & Intelligence', McGraw-Hill International, 395 - 405.
6. Ming-Kuei Hu, 1962, 'Visual Pattern Recognition by Moment Invariants', IRE Transactions of Information Theory, IT-8, 179 - 187.
7. R-C Luo, W-H Tsai and J C Lin, 1984, 'Object Recognition with Combined Tactile and Visual Information', Proc. 4th Conference on Robot Vision, 183 - 196.
8. H Asada and M Brady, 1986, 'The Curvature Primal Sketch', IEEE Transactions on pattern analysis and machine intelligence, PAMI-8, no 1, 2 - 14.

Chapter 16

Stereo vision for flexible assembly —geometric techniques

A. K. Forrest

1. Introduction

To automate most production good feedback from the world is an absolute necessity. Micro-switches are only sufficient in a minority of production areas where many thousands of each component is made. Well over half of production is done in batches of twenty or less. To automate these cases on a case by case basis is obviously not practical. Good feedback from the environment and 'intelligent' use of information is required.

Sight is the sense people use most, and any assembly process can be performned using sight and touch. My group has been concentrating on re-creating a usable sense of sight for computers.

Humans can infer depth from single images using past experience and 'common sense'. Common sense is probably the hardest problem in artificial intelligence. An alternative more suited to the current state of computing machines is needed. Stereo is just such a technique.

Stereo is really just a process of triangulation from two points in space. This is the kind of problem at which current computers are very good. Complex floating point operations can be accomplished at very high speed. It was therefore decided to adopt this approach to producing an accurate three dimensional map of the production area. We have been backed by the SERC and Westland Helicopters to do this research. Helicopters are very expensive and are made in small batches by very skilled and hence expensive workers. They also have the aerospace requirement of tracking every part from batch of metal to final part, for error free quality control. The cost of vision systems in this environment is reasonable and the assembly operation is already tracked by sophisticated computing. Unfortunately Westlands was taken over and dropped this research in the UK, however this particular project has been completed.

2. Technical problems

Although triangulation is a well understood process there are two major areas of difficulty in stereo vision. The first is that to perform triangulation one must know exactly where one is triangulating from, and the second is that the object of interest must be seperable and recognizeable from the two or more positions from which triangulation is to be made. The first problem has largely been solved by ourselves and others. The second still requires considerable work for a completely general purpose system, although in limited situations the problems in this second area can also be solved.

3. Camera calibration

This is effectively the process of finding out where the camera is, which direction it is facing and the camera characteristics, such as the focal length of the lens. This can be done by straight forward measurement, for example by putting the camera at the end of a robot arm. Position and angle are known to the accuracy of the robot. In practice this method is not very useful. It is time consuming and complex to make the measurements on the camera and any errors in these measurements or in the robot position directly produce errors in the stereo data.

A much better technique is to put a known object in the field of view of the camera. As long as the object has six recognizeable points one can obtain the camera calibration matrix. That is the transformation between real world x,y and z coordinates and u,v image coordinates. This matrix is of size 4x4. Twelve of the terms can be calculated from the known object and the remaining four terms can be calculated from these twelve.

We have extended the idea to take account of camera non-linearity and to produce a best fit calibration under the known constraints.

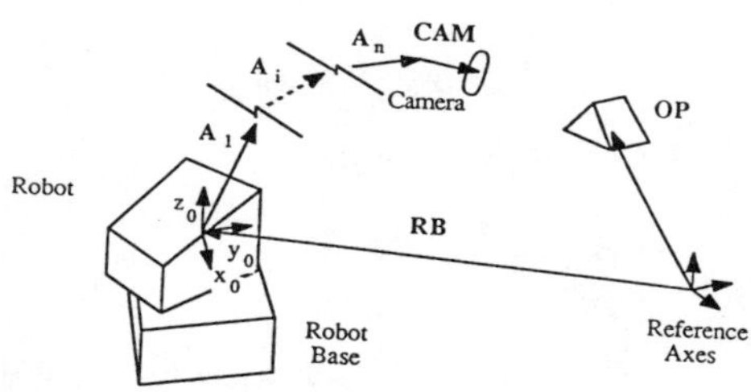

FIGURE 1 Physical basis for mathematics.

4. Camera linearity correction

Let us define three sets of image coordinates $[j, k]$ the perfect image, $[u, v]$ the actual image, and $[j', k']$ the corrected image converted from $[u, v]$.

IDEAL IMAGE			OBSERVED IMAGE	CORRECTED IMAGE	
$[j, k]$	distortion	$\Rightarrow$	$[u, v]$	correction $\Rightarrow$	$[j', k']$

Let us further assume the form of the distortion as follows :-

$$j' = a_0 + a_1 u + a_2 v + a_3 u^2 + a_4 uv + a_5 v^2 \tag{1}$$

$$k' = b_0 + b_1 u + b_2 v + b_3 u^2 + b_4 uv + b_5 v^2 \tag{2}$$

and define :-

$$A = [a_0, a_1, a_2, a_3, a_4, a_5]^T \tag{3}$$

$$B = [b_0, b_1, b_2, b_3, b_4, b_5]^T \tag{4}$$

We want to minimise the difference between the ideal $[j, k]$ and the corrected image $[j', k']$ that is:-

$$E = \sum (j - j')^2 + (k - k')^2 \tag{5}$$

$$= \| j - PA \|^2 + \| k - PA \|^2 \tag{6}$$

where:-

$$P = \begin{pmatrix} 1 & u_1 & v_1 & {u_1}^2 & u_1 v_1 & {v_1}^2 \\ 1 & u_2 & v_2 & {u_2}^2 & u_2 v_2 & {v_2}^2 \\ \vdots & \vdots & \vdots & \vdots & \vdots & \vdots \\ 1 & u_i & v_i & {u_i}^2 & u_i v_i & {v_i}^2 \end{pmatrix} \tag{7}$$

We want to calculate the vectors A, B. We know that

$$j = PA \tag{8}$$

By re-arrangement of this equation we obtain

$$A = (P^T P)^{-1} P^T j \tag{8}$$

$$B = (P^T P)^{-1} P^T k \tag{9}$$

We know the values of j, k ,they are our test object or chart, and we know P these are the image coordinates, hence we can obtain the non-linearity vectors A, B. The only practical difficulty is to arrange that there is no unintended rotation betwen the test chart and the camera which will give false results. This process allows us to remove any small camera spatial non-linearity. Accurate camera calibration requires sub-pixel accuracy in the image coordinates so care on image linearity is essential. Full details are given in Bowman and Forrest (2).

5. Camera calibration 'best fit'

The method of using a calibration object to obtain the camera matrix works well, however each term in the camera matrix is treated as being independent. This is not the case. We know the exact form of the camera matrix which is composed of rotations, shifts, scalings etc. It is therefore possible to obtain a best fit camera matrix using the known form of the camera matrix as a constraint. This can be done using a complex multi-dimensional iterative minimisation scheme. This however is time consuming. We have found that by extracting the rotational part of the matrix, which is 3x3 we can use the property of orthogonality of rotation matrices as the basis for a non-iterative process.

The process depends on the following identities where N is an anti-symetric matrix and R is an orthogonal matrix.

$$N = (I + R)^{-1}(I - R) \tag{10}$$

$$R = (I + N)^{-1}(I - N) \tag{11}$$

Using the equation (10) we can obtain an almost antisymmetric matrix from the almost orthogonal rotation matrix.

$$N' = \begin{pmatrix} e_{11} & l + e_{12} & -m + e_{13} \\ -l + e_{21} & e_{22} & n + e_{23} \\ m + e_{31} & -n + e_{32} & e_{33} \end{pmatrix} \tag{12}$$

We can work out the expected values of l m and n by taking the averages of the terms ie.

$$l = (N'_{12} - N'_{21})/2 \tag{13}$$

$$m = (N'_{13} - N'_{31})/2 \tag{14}$$

$$n = (N'_{23} - N'_{32})/2 \tag{15}$$

$$N'_{11} = N'_{22} = N'_{33} = 0 \tag{16}$$

We can then convert back to a now perfectly orthogonal rotation matrix using identity (10). There is one practical problem $(I + N)$ is never singular but $(I + R)$ can be. This problem can be overcome simply by operating a ninety degree rotation matrix on the subject matrix , performing the calculations on the now non-singular matrix, and then transforming back.

6. Tracking

From the camera calibration matrix it is possible to obtain the relative position and angle of the camera to the calibration object. It is therefore possible to track a calibration object or any object with six or more recognizeable points whose relative positions are known. It is however more accurate to use a differential technique. The imaging process is described below.

$$CAM.X.OP = IP \tag{17}$$

$$CAM.X'.OP = IP' \tag{18}$$

Where CAM is the camera matrix, X is the position of the camera, OP is the object points vector, and IP is the image points vector. Equation (18) is the case after a small movement of the camera. In the above case the characterstics of the camera (ie. focal length,scales) have been split off into the matrix CAM and the position and orientation part is in matrix X. $\delta x, \delta y$ and δz are rotations about the three respective axes, S_x, S_y and S_z are shifts.

$$\delta X = \begin{pmatrix} 1 & -\delta z & \delta y & S_x \\ \delta z & 1 & -\delta x & S_y \\ -\delta y & \delta x & 1 & S_z \\ 0 & 0 & 0 & 1 \end{pmatrix} \tag{19}$$

$$X' =^0 \delta X.X \tag{20}$$

$$X' = X.^X\delta X \tag{21}$$

${}^0\delta X$ is with respect to the camera, and ${}^X\delta X$ is with respect to the object. By substituting $u_i' = u_i + \Delta u$ and $v_i' = v_i + \Delta v$ in equation (22) ,where i is for each image point, one can obtain $\delta x, \delta y, \delta z, S_x, S_y, S_z$ in terms of u_i, v_i and $\Delta u, \Delta v$. If there are at least three image points and second order terms are neglected the components of δX can be calculated.

$$CAM.^0\delta X.X.OP = IP' \tag{22}$$

TABLE 1 Resultant errorors of tracking algorithms

Shifts			Direct		Differential	
x	y	z	Δ Trans	Δ Rot	Δ Trans	Δ Rot
0	10	0	3	0.1	0.35	3.9
10	10	0	2.5	4	0.86	2.1
0	100	0	12	17	11	33

Distances are in mm. rotations in milli-radians.

The results in table 1 are from a practical test where the test object was moved a known amount. The total distances moved are $10, 10\sqrt{2}, 100$ millimetres.

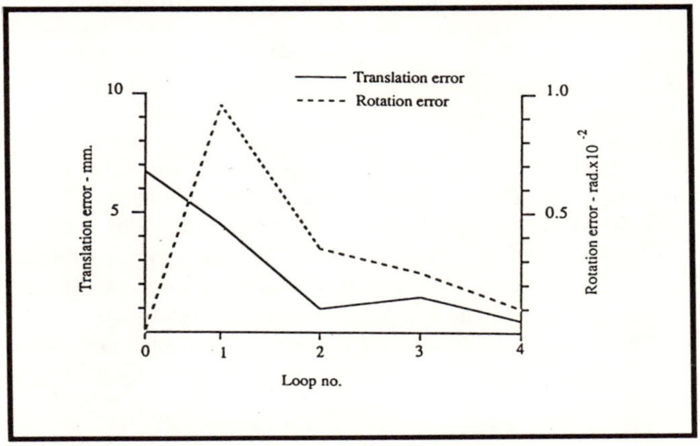

FIGURE 2 – **Top** ,Convergence of tracking algorithm.

FIGURE 3 – **Bottom** ,Typical stereo input image.

The differential technique is, surprisingly, more accurate for small movements than the absolute solution. This is probably due to the techniques non-reliance on the details and accuracy of the robot geometrical model, see Bowman and Forrest (2).

7. Robot calibration

Having put all this effort into finding where the camera is in relation to a known object we can use the method to test the robot's control algorithm. Robot accuracy drifts away from the repeatability because the mathematical model of the robot in the control system no longer corresponds with the actual case. With independent information on the robot effector position from the vision system we can generate a new and more accurate robot model. Again we can model the imaging process but in this case we explicitly include the effect of each robot joint by use of the well known 'A' matrices. RB is the transformation between robot coordinate system and that defined by the reference object $A_1 ... A_n$ are the position transformations for each joint.

$$CAM.(A_1...A_n)^{-1}RB^{-1}.OP = IP \tag{23}$$

To succeed in this re-modeling we must know or find all free variables in the system. There are several unknowns concerned with the exact details of the camera mounting. The complete list of variables is as follows

$\Theta_{1...n}$	Joint offset angles
P_{xr}, P_{yr}, P_{zr}	Offsets from robot coordinate system to calibration object coordinate system.
$\alpha_r, \beta_r, \gamma_r$	Rotations between robot coordinate system and calibration object coordinate system
p_{xc}, p_{yc}, p_{zc}	Translations from joint A_n to the camera lens.
$\alpha_c, \beta_c, \gamma_c$	Rotations from joint A_n to the camera lens.

$CAMM$ is the camera matrix including The above unknowns. The position of the robot is taken from two sources, the robot control system (equation (24)) and the camera calibration system (equation (25)).

$$X = Translation_{(P_x,P_y,P_z)} Rotation_{(x,\alpha)} Rotation_{(y,\beta)} Rotation_{(z,\gamma)} \tag{24}$$

$$X' = CAMM.(RB.A_1...A_n)^{-1} \tag{25}$$

$$E_{model} = \sum \| x_i - x_i' \|^2 \tag{26}$$

The error in the model,E_{model} is shown in (26) where x_i are the components of matrix X. In our practical system we used ten views of a calibration object with twelve known points in each. A multi-dimensional minimisation routine is necessary to find the best fit robot parameters. In our case this is a sophisticated routine based on Davidon's variable metric method with modifications but convergence still requires approximately fifteen minutes on a microVAXII.

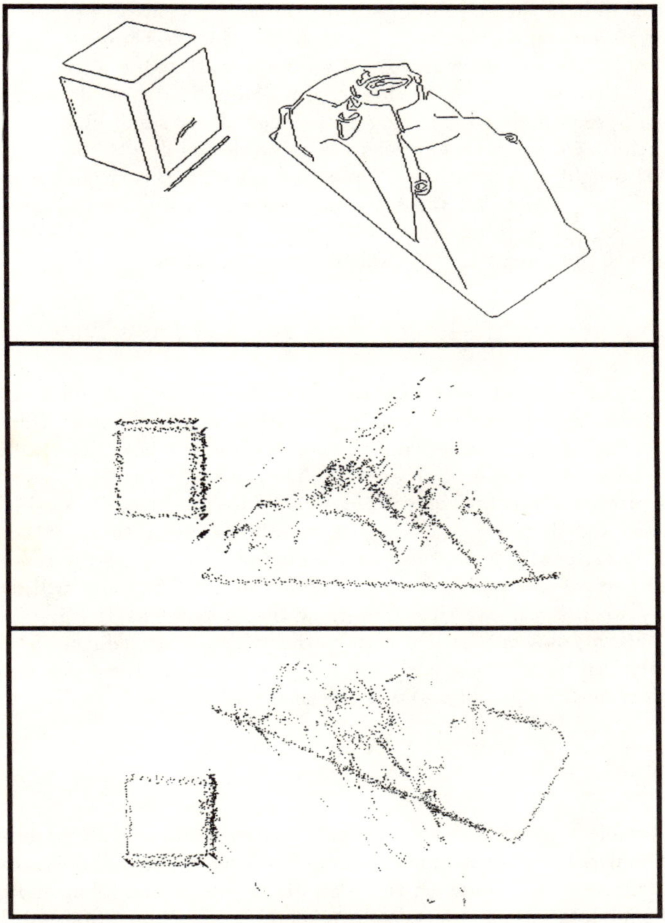

FIGURE 4

Top – Typical processed image used by the routine.

middle – Three dimensional data obtained projected on z=0 plane.

Bottom – Three dimensional data obtained projected on y=0 plane.

8. The combination of techniques

When combined these techniques can completely control a robot. The robot does not need to have great accuracy, especially if visual tracking and position measurement is used. The present high cost of robots is mainly due to their requirement for accuracy. If the robot is used in close conjuction with a three dimensional vision system accuracy requirements can be greatly relaxed. A small change in accuracy requirements can have a drastic effect on the robot design due to the circular dependence of robot weight and accuracy. When controlled by the vision system there is the considerable advantage that the coordinate system used is that of the calibration object ie. that of the working area, not a robot relative coordinate system. This greatly reduces the problems of robot installation and setup.

9. Stereo

Having complete control and knowledge of the camera position the remaining problem for stereo is matching between images. We use a three view technique with all views taken by one moveable camera mounted on a robot. The robot will be necessary for the assembly operation anyway. There are two geometric constraints on the matching process. First the three dimensional data produced by a match must be consistent across the three views. This is the reason for using three, rather than the more normal, two views. The second is epipolar lines. It must be possible to put a plane through two camera positions and an object point. The point will lie on a line in each of the two images. We therefore know that a point on the line in one view must match with a point on the line in the other view. This reduces the number of matches greatly. An input image is shown in figure 1 and the three dimensional data obtained project on the x,y and z axis is shown in figure 2.

10. Conclusion

It is possible to have a robot assembly system completely controlled from a three dimensional vision system. This includes tracking in three dimensions, controlling the robot movement , producing the computer model of the robot and producing three dimensional data on the objects to be assembled. The limitations on the system at this time are the cost of the necessary computing power, and the ability to work in highly un-constrained environments. The ability to work in very unconstrained environments needs more research. The cost of computing power is coming down rapidly. In the future the most cost-effective method of performing flexible assembly will be under the control of a three dimensional vision system and even today this is true in some environments.

11. References

1 – M.E.Bowman and A.K.Forrest 1987, The transformation Calibration of a camera mounted on a Robot,**Image Vision and Computing**, Vol 5, no 4, 261-266.

2 – M.E.Bowman and A.K.Forrest 1988,Visual detection of differential movement: Application to Robotics, **Robotica**,vol 6, 7-12.

3 – Ito Minouru Ishii Akira 1986, Range and Shape measurement using three view stereo analysis, **IEEE Computer Society Proc. Conf. Computer Vision and pattern recognition**

4 – Ito Minouru and Ishii Akira 1986, Three view Stereo Analysis,**IEEE Trans. on Pattern Analysis and Machine Intelligence**, No.4.

5 – R. M. Haralick and Y. H. Chu 1980, Solving camera parameters from the perspective projection of a parameterized curve, **Pattern Recognition**,Vol 17, No.6.

6 – I. Sobel 1974, On calibrating computer controlled cameras for perceiving 3-D scenes, **Artificial Intelligence**, Vol 5.

7 – A. K. Forrest 1986, Stereo vision for flexible assembly, **ACME Grantees Conference**.

8 – A. K. Forrest 1988, Stereo vision for flexible assembly-II, **ACME Grantees Conference**.

9 – R. Y. Tsai 1986, An Efficient and Accurate Camera Calibration Technique for 3D Machine Vision, **IEEE Computer Society Proc. Conf. Computer vision and Pattern Recognition**

Chapter 17

Research into on-line correction and adaptation of automated manufacturing processes using inspection and quality control data

D. C. Robinson and J. Billingsley

SUMMARY

This paper describes a system which has been designed to help improve the competitive position of small to medium size manufacturing companies. It is a system which can be added to the production line to increase quality and reduce costs by exploiting latest technologies. The system is not a substitute for full automation, but it can be used for existing production lines with almost the same results. This makes it very inexpensive and easy to introduce.

For the purposes of this paper a case study is presented. The company is Turnright Controls Ltd., a manufacturer of controls for the domestic appliance industry. The selected product is the Energy regulator.

Introduction

Over the last few years manufacturing has changed dramatically. Information technology is no longer limited to the office environment. Computers are emerging on the shop floor to assist in manufacturing, management and control. More recently computer technology has been exploited to assist in quality control.

In many companies control is exercised through production engineering from verbal or written reports from quality assurance. This has the effect of introducing time lags between failures and corrective action, which can be quite costly. Any statistical quality control is usually performed manually at key stages using instruments such as control charts. The use of computers can greatly improve the quality performance in manufacturing and reduce the aforementioned costs. Unfortunately, commercially available systems are difficult to justify in cost terms and are quite basic in terms of the quality control which could be exercised.

At Portsmouth Polytechnic, investigations are in progress to develop techniques to integrate computer assisted quality in production to enhance product quality. As resulting systems are aimed at smaller companies they have to show significant benefits, be cost effective and easily maintained. The work is sponsored by the Science and Engineering Research Council (Application of Computers to Manufacturing Engineering Directorate) in collaboration with local companies.

Commercially available systems

On-line computer quality control packages have started to enter the market, all of which are aimed at larger companies. Data acquisition terminals, automatic or manual, are located at the desired processes. Data is collated and fed into a computer. Process control charts can then be displayed to an operator, off-line. From this the operator makes a decision as to whether corrective action at a particular process is required. This is technically on-line, postprocess control with some delay between failures and action. The advantages of such systems are that corrective action can be taken earlier and that subjectivity in the acquisition of data is largely removed. One in particular has the further advantage that it can be integrated with all other functions within a company, and this particular vendor provides other systems for these functions.

The main disadvantage with such systems is their cost, which is unjustifiable to smaller companies. Another limitation with these systems is the lack of diagnostics. Most have databases, but none use this data to determine the nature of the fault in order to save time in corrective action. The systems merely inform the operator that there is a problem, leaving the analysis and degree of rectification to the engineers. Therefore there is a requirement for maintenance engineers to be present in case of problems on the line. Another limitation is that no on-line control is possible with these systems.

Overall Strategy

All processes can be said to fall into two categories:
1. Variable processes - ones in which a parameter can be adjusted to affect product performance.
2. Constant (non-variable) processes - ones in which there is no variability in the way the process is performed.

Provocative experimentation techniques can be used to determine the mathematical relationship between variable process parameters and product performance. Once the relationships have been determined a model of the manufacturing line can be constructed. All the variable processes are represented in a matrix. Reducing the number of dimensions using multivariate techniques enables all process variables and their relative effect to be represented by linear transformations. This enables the computer to determine parameter correction in response to quality variations. The overall aim is to continually monitor test and process data, ensuring the correct process parameters to maximise quality performance. At the same time the constant processes can be monitored and helpful diagnostic information on their performance can be fed-back to reduce the possibility of rejects being produced. Thus as there are two types of processes there are also two types of feedback:
1. Direct Feedback (variable processes). Using multivariate analysis of variations techniques, coupled with optimisation techniques, the parameters of each process can be adjusted to values which together produce an optimum product performance. This removes the problems associated with islands of automation, produced by currently available commercial systems.
2. Indirect feedback (constant processes). Diagnostic information from the system can be supplied to production personnel and used to increase awareness of shop floor performance and highlight possible problems.

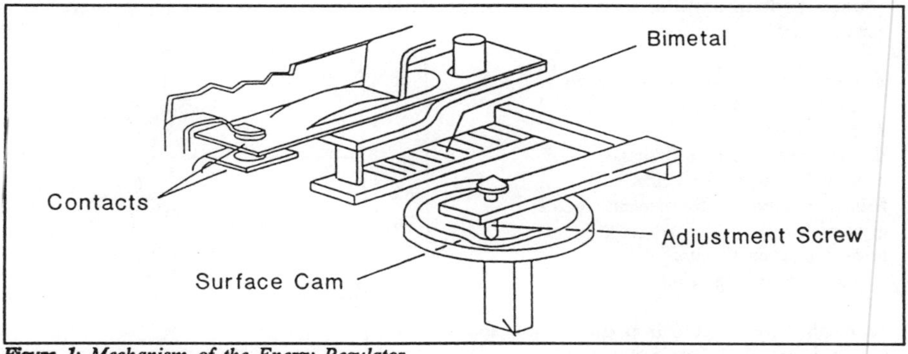

***Figure 1:** Mechanism of the Energy Regulator*

Computer assisted quality and the manufacture of energy regulators

The main advantage in introducing computer assisted quality control into the manufacture of energy regulators is to reduce quality costs. These costs can be divided into tangible and intangible costs. The tangible costs are those associated with rejects, rework and inspection. The intangible costs are those associated with the loss of business due to defect products being dispatched to customers. Our approach is to introduce control to all of the up-stream operations as required, from the results of the final test. In some cases the task is fairly straightforward, with linear relationships between error and parameter correction. In other cases it is difficult to distinguish the operations or parameters which caused the error. Patterns of data have to be associated with particular causes of error. This involves the creation of a database with test, operational and failure information.

The Energy Regulator

The Energy Regulator is a control device which cycles power to heating elements on electric cookers. Five such devices are usually present in each cooker. See figure 1.

It consists of a bimetallic strip which is heated in parallel with the heating element on the cooker. As the strip is heated it bends until it forces a copper spring to snap open a pair of contacts. This then interrupts the supply, and the bimetal cools. As it cools, the copper switch is again bent until it snaps closed the contacts, restoring the supply. The amount to which the copper switch has to be bent to change contact state is a measure of the hysteresis of the device. This in turn is the main determinant of the mark-space ratio of the Energy Regulator. The mark-space ratio is the rate at which power is switched to the heating elements for a certain setting of the control knob. The control knob on the cooker is located on the spindle of the regulator. An adjustment screw is incorporated to offset any build-up in deviations as a result of manufacturing operations.

Manufacture

The energy regulator is a high volume product with an annual output from the factory measured in millions. The mechanism of the device is fairly simple, whereas the quality problems arising from production are quite various.

Manufacturing quality control begins at Goods Inwards where sample inspection is employed to determine the suitability of batches of components. Some components are produced on site, which facilitates adjustments to these processes if unsatisfactory batches are produced. Manufacture is then separated into three main sub-assemblies:

1. Base assembly; comprising the housing, spindle, cam, cam spring and base contacts.
2. Heater assembly; comprising the bimetallic strip, the heater (a coil of resistance wire), pivot and adjustment screw assembly.
3. Carrier assembly; comprising the metal cover and the switch assembly.

Involved in the assemblies are many process elements each of which can introduce irregularities in product performance. For example the processes involved in producing a heater assembly are briefly detailed below:

a. receive components and arrange
b. cleat bimetallic strip to board
c. spot weld
d. wind heater coil on bimetal
e. weld heater to pivot arm
f. heat in oven to stress release welds
g. varnish
h. heat in oven to cure
i. 100% visual inspection

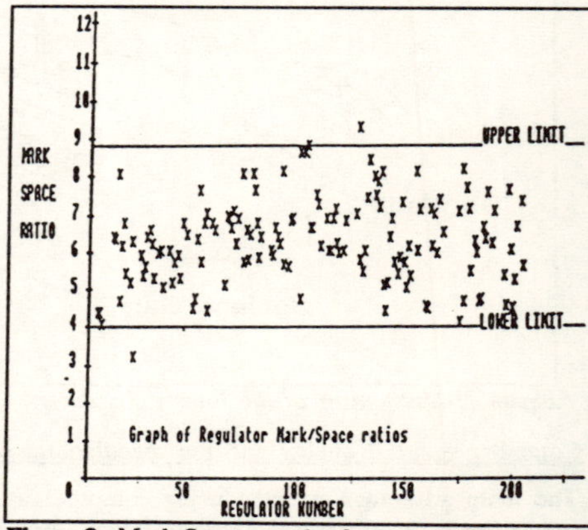

Figure 2: Mark-Space results for one batch

Although there is 100% inspection at this stage it is only visual defects which can be spotted. Moreover, studies have shown that 100% manual inspection is only a maximum of 85% effective, depending on the nature of the task. Other defects cannot be found until after final assembly.

After the sub-assemblies have been completed they undergo final assembly. Here the assembling, staking and pressing functions are carried out.

The Backstop

After the regulators have been assembled, a fixture is employed to bend a small piece of metal in the carrier assembly to a pre-determined angle. This part of the assembly is termed the 'backstop' and has the effect of pre-stressing the copper switch, thus affecting the hysteresis of the device. The resulting gap between the contacts is then checked with a slip gauge. If the gap is incorrect, the operator bends the backstop until the gap is satisfactory. After this stage the regulators are subjected to 288 volts (110% of mains) for five minutes to check for loose wires.

The adjustment screw

The adjustment screw was incorporated to compensate for any build-up in performance deviation. The switching action of the regulator is related to spindle position by way of the adjustment screw riding on the surface cam mounted on the spindle.
The adjustment process is the only process in the line which was continually monitored as a quality controlled process.
Regulators arrive at the adjustment stage in the 'off' position, with the contacts open. The spindle is rotated until the contacts close. The angle of the spindle, relative to the off position at this point is called the 'closing angle'. As the mark-space ratio cannot be measured during adjustment, the closing angle is used as the basis for adjustment. The regulator screw is rotated until the correct closing angle is achieved. It is assumed that this setting will produce a satisfactory mark-space ratio.

The Mark-Space Test

After adjustment the Regulators undergo final testing. This is the definitive test of Regulator quality. Regulators are placed in an array of test positions in test boxes. The spindles are then set to a pre-determined angle, which is expected to produce a particular cycling rate. They are then connected across a regulated mains supply. The times for switching on and off are recorded, from which the mark-space ratio (cycling rate) can be calculated. See figure 2.

Mark-space ratio (%) = (T3-T2)*100/(T3-T1)

where: T1 = time of the first opening
T2 = time of first closing
T3 = time of second opening

and : T3-T1 = cycle time

Manual Adjustment

Manual adjustment practice was to adjust the screw to a nominal closing angle that production engineering had calculated.
Tolerance on this adjustment was always one degree either side of this value. If it was discovered that a higher reject rate was being produced, then production engineering would modify the value of the required closing angle. Figure 3 shows the feedback loop for manual corrective action.

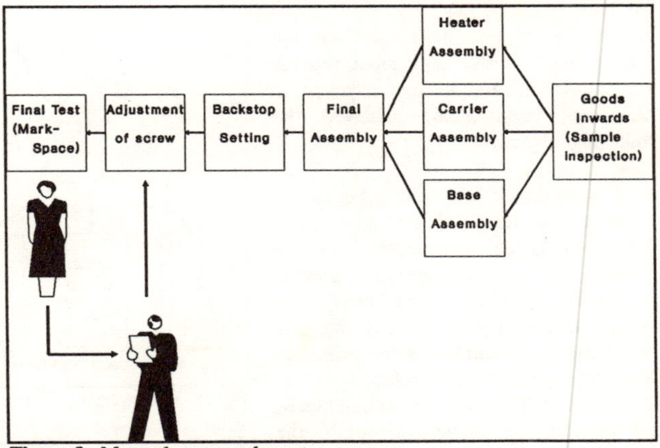

Figure 3: Manual process layout

The problems with this method were:

1. Modifying the required closing angle had the effect of shifting the mean of the mark-space dispersion. Although this had some positive effect in centring the dispersion (moving the distribution curve up or down), there was no control over the width of the dispersion. Modification of the tolerance to which the closing angle was adjusted would have some effect on the width of the dispersion. With a tighter tolerance, the adjustment of the screw would have produced a more consistent closing angle, narrowing the dispersion and therefore producing less failures. This would be at the cost of throughput, as it takes longer to adjust the screw to a tighter tolerance.
2. Rejects had to be produced before any corrective action was taken.
3. Buffer stocks existed between adjustment and final test. By the time warnings were given, batches had been produced which exhibited an abnormal reject rate.

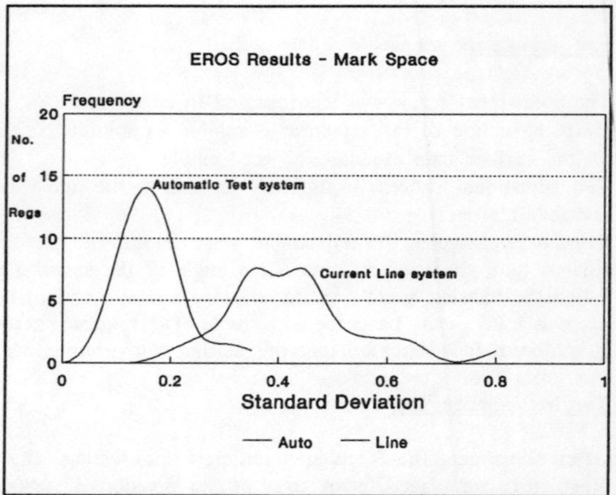

Figure 4: *Graph of relative precision of inspection methods*

Automation of the Final Test

The first stage in integrating automated quality control into the manufacture of regulators was to automate the mark-space test. Regulators are manually loaded into test boxes and when the box doors are closed, mains is connected across them. A microcomputer times the switching of the Regulators and calculates the mark-space ratio. From these ratios, statistical analysis can be used to facilitate diagnosis and to locate trends for preventative control. Automation increased the precision of the final test (see figure 4). This meant that on average the test error was smaller, and therefore less filtering of the data was necessary. It also provided a bonus; the tolerances could be relaxed as the reliability had increased. This meant that the throughput was greater for the same quality levels. In practice the quality was increased and the throughput was increased.

Automation of the adjustment process

At the same time as automating the mark-space test the adjustment process was automated and located adjacent to the test system. This achieved the following objectives:

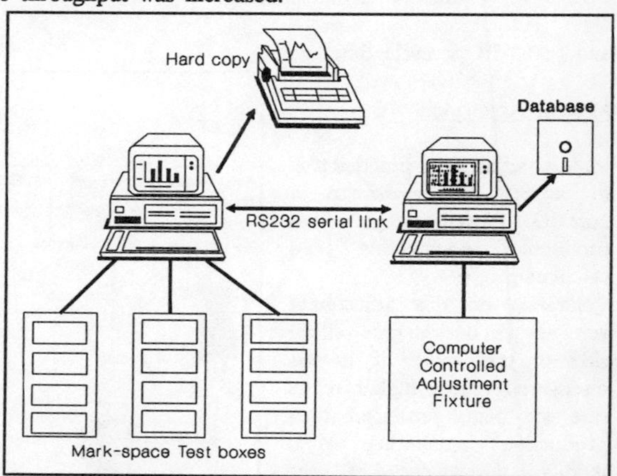

Figure 5: *Automatic Test and Adjustment system*

1. The feedback loop could be closed between the test and the adjustment. This facilitates automatic on-line correction of adjustment process parameters. After adjustment the regulators are loaded directly into the test boxes. This eliminates the problems associated with buffer stocks.
2. The spindle could be automatically set to the test angle. Previously the angle was set manually and its value differed from operator to operator, reducing the reliability of the

test. The angle is now set automatically after adjustment, and the setting is verified by a template and optical sensor.
3. Valuable information about each regulator such as the angular hysteresis could be collated.
4. More accurate and reliable adjustment could be achieved.
To close the feedback loop between final test and the adjustment process an RS232 serial link was used. See figure 5.
Test information is relayed to the adjustment computer from the test computer. This information is used to continually monitor the capability of the adjustment process.
The adjustment of the required closing angle is derived from the test results by altering the target closing angle in proportion to the mean value of the mark-space test from its ideal value, that is:

NEW CA = OLD CA + (-Ka)*(Xms-Xideal)

The value, Ka was determined from the relationship between mark-space and closing angle setting. Although there is a definite relationship, the mark-space performance is a function of many other variables.

The Results

When the defect rate increased under manual operation, the only corrective action taken was to change the target closing angle. This had the effect of moving the distribution curve laterally. The width of the distribution changed with process variations, from batch to batch. Process variations meant that the capability of the manufacturing line was varying. Thus, changing the target closing angle had some effect on centring the dispersion, but the process could still have been out of control. Figure 6 shows the mark-space data for a batch of 250 regulators which had been adjusted automatically. A low sample size of 8 test results was chosen to show the effect of changing the closing angle setting. After 8 results have been received, the parameters of the adjustment process are reviewed and any changes made. It can be seen from the graph that correcting closing angle parameters alone does not always produce the desired effect, suggesting that adjusting only one process is inadequate.
The automatic adjustment system also incorporates changes in the tolerance to which the closing angle is set, in response to the dispersion of the test results. This attempts to keep the process capability at six standard deviations for the mark-space results. The new tolerance is calculated as a product of the old value with the ratio of bandwidth to six standard deviations, that is:

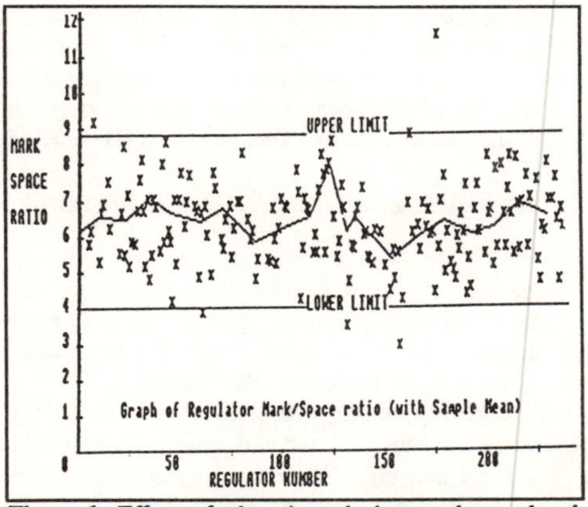

***Figure 6:** Effect of changing closing angle on batch mean*

$$\text{NEW TOL} = \frac{\text{Kt* BANDWIDTH * OLD TOL}}{6 * \text{SD}}$$

Where the gain, Kt was determined experimentally

So, as the standard deviation of the test results change, the adjustment process tries to compensate. With the limits set at six standard deviations, we would expect 0.3% failures, but in practice it was found that altering the parameters of the adjustment process yielded only partial success. This is shown in figure 7.
Although the closing angle is relatively constant due to adjustment of the screw, the hysteresis of the regulators will vary quite considerably. The angle to which the backstop is set has an effect on the

hysteresis of the regulators. Therefore test data should be used to modify this angle. Although there is some correlation between the hysteresis and the mark-space results, there are discrepancies suggesting that hysteresis is not the sole determinant of the mark-space performance of regulators.

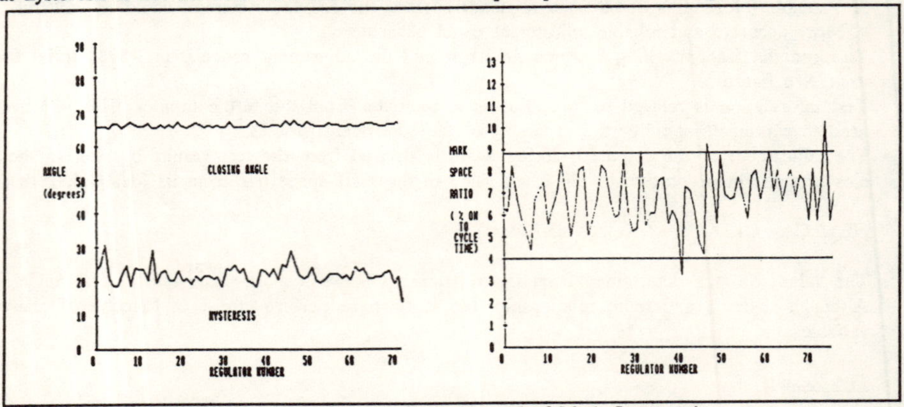

Figure 7: Graph of closing angle setting with hysteresis and of Mark_Space ratio

Stability

During adjustment the response of the device to adjustment is recorded. This data has revealed a new feature in determining regulator quality. It has been discovered that some regulators exhibit variations in performance when tested for a second time. Thus, regulator stability could be an important criterion for determining quality. It is not feasible to multiple test each regulator, therefore stability had to be ignored and was assumed to be a function of the test error. Now with automation, regulators can be assessed in terms of stability criteria from data collated at one or more of the variable processes. The reaction of the product at one of these stages is measured and subsequent analysis determines product performance stability. This increases the reliability of the product without reducing product throughput.

Application of the overall research strategy to the energy regulator line

To illustrate the methodology, consider the energy regulator line. For quality control purposes the line has been segregated into main areas.

1. Adjustment (of screw) process. This is a directly variable process as the altering of screw position directly affects the performance of the product. The exercise of computer control enables the process to achieve its maximum efficiency.
2. Backstop setting process. Again a variable process with a direct relationship between the angle to which the stop is bent and final product performance.
3. Final assembly. A series of constant processes; the regulators are assembled correctly or incorrectly. Diagnostics can assist in locating problems with fixtures, operators and so forth.
4. Heater assembly. A combination of variable and constant processes. The variables are parameters such as oven temperature and weld current. Other processes are carried out in the correct manner or not.
5. Carrier assembly. A series of constant processes.
6. Base assembly. A series of constant processes.
7. Goods inwards. In this area, components are accepted or rejected and it can therefore be said that this area is a series of constant processes. Sample inspection is employed to inspect the thousands of goods passing through the area each day. This means that there is a possibility of defect batches of components getting through and causing reject goods to be produced. Diagnostic feedback is used to reduce the effect of defective components reaching manufacturing, by locating the defective components using past experience via the system database and correlating.

With the main variable process (adjustment) under computer control, the next stage is to automate as many of the other variable processes as possible, or provide some other means of modifying variable process parameter settings. Once achieved the system can be expertly directed by the computer. See figure 8. The computer will monitor the data received from line processes and from the final test. It then takes action to correct any variations from ideal quality levels. After filtering the data and correlating with the knowledge base, it is possible to assign three types of error:

1. Errors which can be corrected by adjustment of variable processes, for example backstop setting.
2. Errors which could be corrected from the use of diagnostic information supplied to the manufacturing line. For example defective fixturing.
3. Errors which could be corrected by a combination of the above actions. For example a specification change for the bimetallic strip could have the effect of increasing the mark-space cycling of the regulators. Modification of the adjustment process and other variable processes can compensate for this event and reduce the mark-space ratio to acceptable levels. Also, the computer would inform production that a change had occurred in the bimetal, and then goods inwards would be informed. The bimetallic strip would be located as the source of the variation in quality by filtering the data and matching error information with past experience from the database.

If an error occurs which is new to the system, that is there is no discernable data pattern which can be associated with that type of error, then the system asks for an investigation to be made. The resulting data is added to the system database.

The system has been completed in part. It is hoped to commence some testing in the near future and obtain the following enhancements to the energy regulator line:

1. Reduction of the nominal failure rate, currently at 6%. The actual failure rate is higher than this, but regulators are re-tested until the error in measurement is such that they display a pass result.
2. Increase in reliability of the product, with resulting increase in customer satisfaction. Quality is rapidly becoming more important than price.
3. Increase in throughput. Reduction in test error means that the tolerance band can be increased, increasing the throughput.
4. More accurate preventative maintenance from diagnostics.
5. Reduction in quality costs with fewer rejects, re-tests and re-work.
6. Simulation becomes possible.

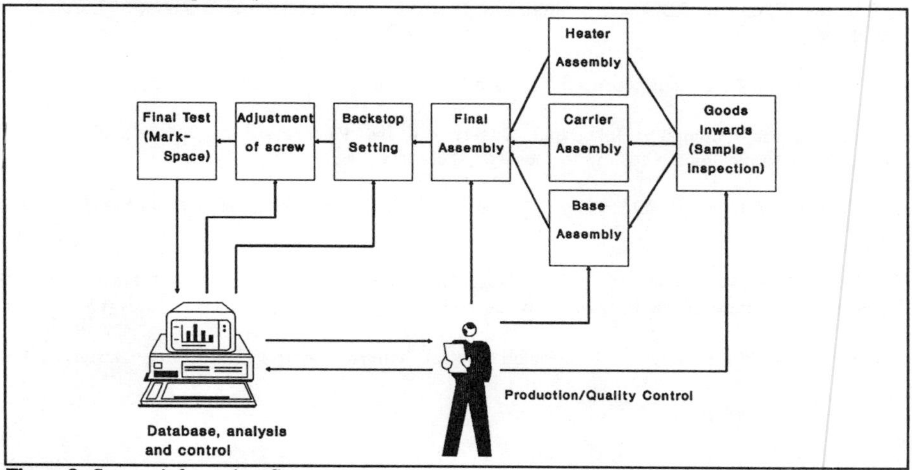

Figure 8: *System information flow*

Simulation

One other very valuable asset of the system is simulation ability. Once every few months, the energy regulator undergoes a design change. Even with careful attention such design changes can affect the performance of the regulator, and sometimes prove to be quite costly. Using the system software model, design changes can be tried and tested before they become reality. This gives greater confidence in product modifications and reduces costs.

Conclusion

With increasing competition, companies are finding that they have to improve quality and reduce costs. The benefits of introducing systems of integrated quality control are that quality costs can be reduced and product quality increased. This paper has shown one example of the implementation of modern quality control techniques, which is aimed at smaller companies. The same techniques could be applied in a large number of companies. The described system is not a substitute for full automation, it is a cost effective method of achieving almost the same results when higher levels of automation can not be justified. It is a method of adding to the existing manufacturing framework in such a way as to enhance the quality control of the line. As it is an addition to the existing processes it is cost effective and very quick to commission.

References

1. J.Billingsley, D.C.Robinson, I.Stott, "On-line correction and adaptation of automated manufacturing processes using inspection and quality control data", ACME Research conference proceedings, Nottingham, September 1988.

2. D.C.Robinson, J.Billingsley, I.Stott, "Computer assisted quality for the small company", International conference on Factory 2000, integrating information and material flow, August 1988.

3. B.L Luk, F.Naghdy, J.Billingsley, "Stochastic Force Sensing Application in Robotics",Euromicro 88, Zurich, August 1988.

4. J. Hollingum, "Craftsman concept achieves a better product", The Industrial Robot, June 1987.

5. D.C.Robinson, "Computer Integrated Quality and the Factory of the Future", digest of IEE colloquium, 'Automation - the way to Quality?', February 1988.

6. F.Naghdy, J.Billingsley, D.Harrison, "Simulation of human judgement in a robot-based adjustment system", Robotica, Volume 2, 1984.

7. D.C.Robinson, J.Billingsley et al., "Craftsman Computer Integrated Quality and Manufacture", 2nd International Conference on Robotics and the Factory of the Future, San Diego, July 1987.

8. G.J.Blickley, "Statistics Invade Distributed Control Systems", Control Engineering, November 1987.

Chapter 18

Management on a collaborative advanced research project

W. J. Finlay

1 INTRODUCTION

In an increasingly competitive world market, centres of research and development are under pressure to design better quality, higher function and lower cost products. One result of their success in this endeavour is higher demand for the product and often, a reduced product life cycle as the next generation of products emerge from development. An enlightened approach to "design for manufacture" goes some way to alleviate the problems experienced by the manufacturing industry in this fast changing environment, but the impact is nonetheless significant. Shorter product life cycles and higher volume production demands can have a severe cost impact in terms of process layout and re-tooling. Production engineers design their production processes using off the shelf technologies and well tried techniques. There is little time for production process development in many companies and much valuable new knowledge and many potentially valuable techniques from universities and other research institutions can take several years to reach the engineer. Some tuning of this knowledge transfer process between the universities and industry could effectively speed up the availability of such information and provide material of more immediate value to manufacturing. Direct feedback from the manufacturing industry to universities at the appropriate level may thus help in finding effective solutions to some of today's problems and lay the groundwork for solving those of tomorrow.

Collaborative projects between industry and academia offer an opportunity to construct such a link with benefit to both parties. Such ventures however bring with them problems of their own not the least of which is how to plan and manage such a project. Contrasting the short term goal oriented method of operation common to industry with the longer term investigative approach of academia, the idea of managing such a project may seem a daunting task to both academic and industrialist alike. After two years of collaboration on the IBM/Heriot-Watt Project however, experience has shown that it is possible to work in harmony towards a commonly agreed set of goals.

2 THE PROJECT

The project, funded by IBM UK LTD and the ACME Directorate, started in June 1986, its main theme being the investigation of error detection and correction in a multi-robot, multi-sensor assembly cell. The theme recognises the need for improvement in error detection and correction techniques in robotic assembly cells and extends the problem to a search for potential benefits from multi-robot, multi-sensor operation. Using a small hydraulic pump as the target assembly, the project investigates the various potential error classes in the assembly operation, the degree and method of sensing required to detect these errors, the decision mechanisms and knowledge required to generate and execute recovery plans and the software architecture to integrate all of these.

The main hardware in the cell consists of two orthogonally mounted cartesian robots - one IBM 7565 and one Pragma H3000 with a small RTX (modified scara) robot for general purpose use. These act in response to commands from the IBM 6150 processor via RS232 links. The frame/agent software supports virtual sensor input and facilitates the addition of further sensors as required. Visual and tactile sensor input is consolidated into a world model representation of the cell and is regularly updated by a world model manager. The assembly proceeds based on the initial assembly plan modified as necessary by the state of the world model [Herd et al, 1].

Supporting this complex project over its first three year phase is a team of six researchers with complementary backgrounds in electronics, software, control and mechanical engineering and mathematics. Under the general topics of system architecture, errors and the world model, collision avoidance, sensor integration, robot vision and assembly strategies each researcher follows his own line of research. As with any project of such a size, a project plan is essential in measuring actual progress against the short and long term objectives. In the university environment where academic freedom is jealously guarded, an agreed project plan provides a key to managing the project effectively.

3 THE PLAN

In formulating the plan for the project several factors require attention. It will be expected by both IBM and ACME that at the end of the funding period a tangible output will be produced. The nature and magnitude of this output will be determined, amongst other things, by the quality of output from individual researchers and by the quality of management control applied in ensuring the integration of these outputs. In contrast to developing a plan in industry where a great deal of past experience can often be applied to the creation of plans for a new project, the research content of this project has made the estimating of task duration much more unpredictable. Changes in individual task schedules are resisted in the interest of overall integration and the solution most preferred by the project management team is to limit the content of the task to a degree suffient to provide a basic link in the total structure.

Our original plan was formulated jointly by IBM Greenock and the University prior to presenting the proposal to ACME for funding. Attached to the plan was a description of the sub-task activities set at a high enough level to allow the maximum flexibility of operation to the researchers. The plan is reviewed with each research associate and all are encouraged to feed back any suggestions for plan improvements. Over the period of the project to date many useful suggestions have been received and implemented. Requests for changes to the plan are reviewed and acted upon at the appropriate level in the management system set up to support the project.

4 MANAGING THE PROJECT

Three main levels of control operate within the management structure : stragic direction managed by the Management Committee, project control managed by the Technical Committee and technical supervision of the Research Associates (R. A. 's) managed by the Investigators.

The Management Committee meets on a regular quarterly basis and is made up of three managers from the Greenock plant and four members of University staff. The IBM managers are a Functional manager, a Second Level manager and one First Line Engineering Manager, the Liaison Manager. The University members are the Professor of Electrical and Electronic Engineering and two of his staff along with a representative from the Department of Mechanical Engineering. The business of the Management Committee is to review general progress of the project in terms of strategic direction, staffing, capital spend rate and previously agreed objectives. The output from these quarterly meetings generally takes the form of a list of items for action by the IBM Liaison manager and the other members of the Technical Committee. In keeping with the mode of operation of the project as a whole, once agreement has been reached in principle, the method of implementation is left to the implementor.

In addition to acting as a useful self-audit function for the Technical Committee, the wider issues of promoting contact with IBM locations and other UK universities from an information gathering point of view are reviewed. With a three year limit on phase one of the project it is important to avoid as much as possible the tendency to reinvent the wheel! Promotion of the project externally, preparations for the regular nine-monthly ACME review and dissemination of results to industry and academia are examples of the topics discussed. From an internal project management point of view it was beneficial during the earlier stages to monitor staff hiring and capital spending to establish a confidence level against the original plan. With the staffing level now on plan and the budget largely spent the committee now concentrates mainly on the means of disseminating results to the academic and industrial communities.

The Technical Committee, consisting of the IBM First Line manager and the same University staff members as are on the Management Committee, is responsible for the detailed planning and management of the project. The research tasks and their progress is reviewed on a regular basis for consistency against the integration schedule and for publication readiness. Minor changes to the research activity are accommodated at this level if required to keep the overall plan on schedule, while more substantial changes are aired at the Management Committee. Dissemination from the academic point of view consists mainly of the publication of papers in journals and conferences and are largely aimed at an academic audience. In preparing a "publication plan" the committee, in consultation with the researchers, prepares a list of possible titles and a tentative target date is set for each. Visits to other UK universities are made based on a selection by the researchers from a knowledge of their published work. On the Industrial side, the IBM Liaison manager is responsible for arranging visits by engineers from the plant both to tap their experience and to demonstrate techniques in use on the project. Since the production engineer has the best experience in applying various techniques to the solution of specific industrial problems it falls largely on the Liaison manager and the visiting engineers to recognise and extract techniques of potential use in the industrial field.

To help in the task of keeping the project on schedule the Technical Committee establishes a set of checkpoints or "milestones" against which overall progress can be measured. In establishing the checkpoints, the research plan is divided (not arbitrarily) into nine-monthly segments - each of which is positioned one month before the scheduled ACME review. For each checkpoint, a projection was made as to the functional capability of the multi-robot, multi-sensor cell at that point in time, based on an estimated rate of progress. From the definition of the function, a set of sub-objectives is generated for each of the lines of research in order to force some degree of integration and these are given to the researchers. While much emphasis is placed on the integration aspect, the objectives are set at a sufficiently high level to allow maximum flexibility to the researcher. Deviations from the set objectives have been necessary in some instances but in general, integration has been maintained.

On a purely technical level, meetings between the investigators and their R. A. 's take place on a weekly basis. The subject of these meetings varies depending on the various stages of progress in each of the research areas. In some cases, the meetings take the form of a progress check and in others the discussions revolve around new research approaches. In general, the R. A. 's and investigators meet in a joint session with most of the team present. One significant benefit from the joint session approach is that everyone in the team is aware of progress and problems in the other areas. These meetings can be, and often are, quite lively events as each R. A. who feels competent to comment on another area of research is encouraged to do so. Since integration of these different areas is a high priority objective the interchange of ideas is extremely useful in this respect.

5 SUMMARY

Managing a collaborative project is undoubtedly a challenge for everyone involved. From the typical academic researchers point of view, one of the most valued aspects of the job is his ability to work unsupervised on the topic of his choice and diverge at will onto a more interesting subject if the oppurtunity arises. In industry, there is no time for diversions. There is no time to pursue "interesting problems." The objective and the schedule must be met. Within each of these spheres of operation - research and manufacturing, each performs his task to the best of his ability and within the adopted guidelines of the job. In attempting to marry the two there would appear to be a distinct conflict of objectives. With some compromise on both sides, in the interest of UK industry in general, a common understanding is the first step towards a new set of objectives for collaborative ventures. A first essential objective is the establishment of an empathy between the participants.

While the enthusiasm of the team members has been an invaluable benefit to the project, a major contribution to the good working relationship between the academic and industrial partners has been the on-site presence of the manufacturing representative. Once established in the university with his knowledge of plant operations, procedures and terminology, he can effectively act as translater, broker and arbiter between the two. Simple misunderstandings in interpretation which can all too easily become major disagreements can be resolved locally. The industrial representative can also provide useful advice and guidance on manufacturing matters on a day to day basis by participation in the meetings at all levels.

A good working relationship has been established between the plant and the university and some results look promising although the project is not yet complete. The management system in place is operating effectively and will continue to be operated in its present form.

6 ACKNOWLEDGEMENTS

The author would like to thank Prof. G. T. Russell, the other members of the research team, J. McLelland, Technical Programmes Manager (IBM Greenock), J. Ogilvie, New Products and Technical Programmes Manager (IBM Greenock) and the SERC ACME Directorate for their support. Funding was provided jointly by IBM UK LTD and SERC under SERC grant GR/D 69778.

7 REFERENCE

1. Herd,J. T., Duffy,N. D. D., Phillip, G. P., Eccles,N. J., Davidson,A. C., "Control of Multiple Robots in a Sensor Rich Environment." 1989 , IEE Fifth Conf. on Exploitable UK Research for Manufacturing Industry.

Chapter 19

Control of multiple robots in a sensor rich environment

J. T. Herd, N. D. Duffy, G. P. Philip, N. J. Eccles and A. C. Davidson

1 INTRODUCTION

The introduction of robots to the factory production lines of the 1970/80s was considered a significant breakthrough in terms of flexible automation. The robot appeared to have many of the attributes of a human worker coupled with tireless, consistent, and uninterrupted performance. In fact robots were found to be well fitted to only a limited set of applications such as spot welding and spray painting. Such applications did not require a high degree of accuracy and there were obvious advantages in replacing operators who would otherwise have to work in adverse environments. Outwith these areas, a number of problems that have a direct effect on production costs make the justification of robot usage more difficult. Notably, the simplest deviations from the predetermined robot-action and part-presentation plan can either cause the production of scrap or the halting of workcell activity. Furthermore, current generation robots utilise relatively primitive sensors and consequently require very accurate, complex and expensive fixturing for all but the most trivial tasks.

One approach to bringing the benefits of robotic automation to a wider range of applications, is to provide a workcell structure in which a multi-sensor based system [Henderson et al, 1] provides information about the working environment to a set of collaborating robots [Koivo & Bekey, 2] in a Multi-Robot-Multi-Sensor {MRMS} workcell. Such a cell not only reduces product changeover times and reduces fixturing requirements but also allows the system to detect and correct errors because the sensors are able to provide detailed knowledge of ongoing workcell activity.

Traditional robot systems with their computer based controllers normally come with some form of specialised robot control language, for example AML or VAL. These languages are sequential in nature and handle sensory information by reading sensory data as required at specific points in the programme sequence. Response to errors is also limited, since error checks and handlers have to be explicitly coded. In situations where inputs from multiple sensors are required or where there is a need to control multiple robots, the use of these languages leads to complex code with attendant problems of testing and maintenance. Some attempts have been made to overcome these problems by the introduction of extended forms of robot control language such as AML/V, but even these extensions fail to cope with the complexity of coordinating the control of an MRMS workcell. The problem is exacerbated when each robot has a completely different programming language.

The controlling structure of an MRMS workcell must be able to easily integrate an assortment of robots and sensors if it is to be applicable in a variety of manufacturing situations. An approach to this problem is to decouple the major system components from one another [Kent & Albus, 3]. This produces a modular system in which modifications of both the system and workcell configuration are simplified. This modular approach also allows the exploitation of the parallelism inherent in both an MRMS workcell and assembly operations.

2 CELL MECHANICAL STRUCTURE

An MRMS workcell has been established at Heriot-Watt University to investigate solutions to the problems described {see figure 1}. The workcell consists of three commercial robots operating in a shared workspace, an IBM 7565, a DEA Pragma H3000 and a UMI RTX. The 7565 is a cartesian robot having 6 degrees of freedom. It is gantry mounted and operates in a cuboid space which forms the workspace for the other two robots. The Pragma H3000 is a beam mounted cartesian robot with 5 degrees of freedom. Its horizontal telescopic arm projects into the workspace of the 7565. The gripper for the Pragma robot has been developed at Heriot-Watt University and uses a move-lock-grip mechanism to provide a 150mm stroke combined with a high {200N} grip force. This allows the gripper to handle a large variety of parts without the necessity of changing the end-effector during the assembly process. The RTX is an enhanced SCARA robot with 6 degrees of freedom.

Information on the operation of the workcell is obtained from a diverse range of sensors. Video cameras provide multiple views of the workspace. Two are mounted on the gantry of the 7565 and a third is mobile, mounted adjacent to the end-effector of the H3000. A laser scanning system is provided which is able to lay down a line of light at controllable angles and positions within the workcell. The scanning system in conjunction with one of the video cameras is able to provide information on the 3-D shape and spatial

Fig. 1 The workcell.

position of objects. Touch information from the three fingers of the Pragma gripper is provided by novel touch sensors based on PVDF material. A Lord sensor on the wrist provides wrist force and torque measurements for insertion force feedback. A set of strain gauges provide three axes force sensing for the gripper fingers of the 7565 .

3 CONTROL SYSTEM STRUCTURE

Control of the MRMS is carried out by three major subsystems, the robot subsystem, the sensor subsystem, and the error handling mechanism. The system is structured in such a way that the task of executing the desired assembly operations is decoupled from that of acquiring the required sensory data. This separation enables the error handling mechanism to compare the expected results of operations within the cell to the results observed by the sensory subsystem.

The splitting of the control structure into three major subsystems is only the first phase of a process of system decomposition to produce a detailed structure that can be physically implemented. The method of decomposition breaks a complex problem down into a large set of sub-problems, each of which are relatively easy to solve and implement. The co-ordinated solution to these sub-problems produces the resultant system activity.

In the sub-systems described here, these sub-problems are implemented as a set of parallel acting processes called agents. Co-ordination of the agents is achieved by communication through common data areas. Related data in these common areas is grouped in structures called frames [Minsky, 4] which can in turn be partitioned into a hierarchy of simple data elements called slots. The parallel-acting agents are data-driven since they interact only by the reading and writing of frame data. No explicit agent control coupling exists.

Sets of agents are grouped together according to the type of processing that they perform. Similarly, the frames associated with each processing group are collected into common data areas called pools. This data-flow type methodology can be illustrated by using a MASCOT like notation [5] {figs 2,3 and 4}. In the notation used a circle represents a set of agents, a dish represents a set of frames and a directed line represents the flow of data.

The diagram shown in figure 2 shows the operation of the system under non-error conditions. The Plan Execution System {PES} is responsible for robot co-ordination with respect to an assembly plan. It consists of a number of agents grouped by function as the Job Manager and the Job Processing & Interface. The user supplies an object-level assembly plan which is read by the PES and translated into robot commands for specific proprietary controllers. These commands are executed as they are generated.

The World Modeling System {WMS} provides the decoupling between the sensors and the PES. The interface between the two is in the form of a common set of frames forming the World Model {WM}. This is a high level, object indexed description of the workcell which is used by the PES to locate objects as the assembly plan is executed. The WMS runs concurrently with the PES, continuously maintaining the best possible World Model from its sensory inputs.

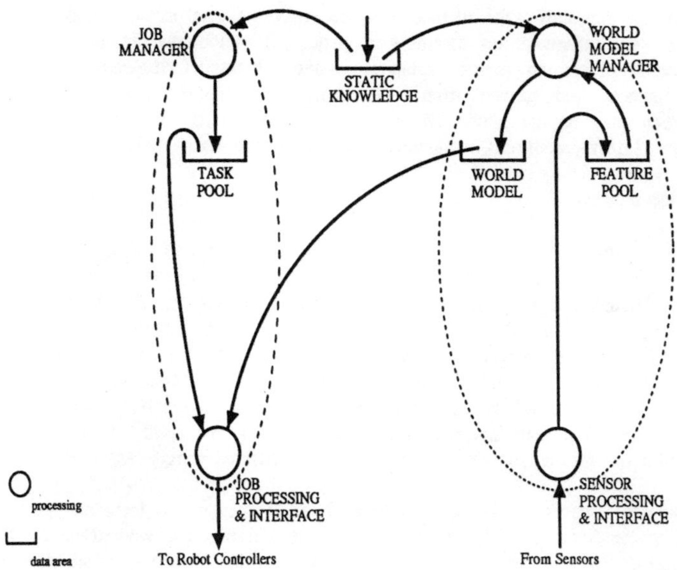

Fig. 2 Subsystem interaction under non-error conditions.

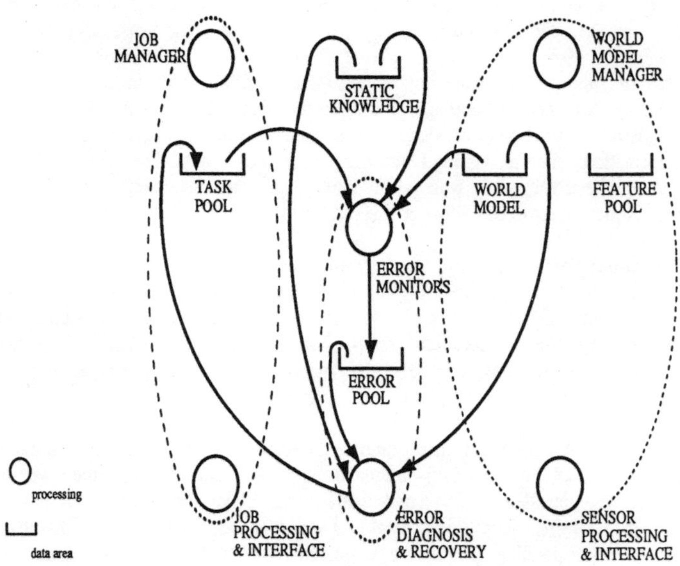

Fig. 3 Subsystem interaction under error conditions.

The Error System allows the workcell to deal with the unexpected events which would otherwise cause the failure of an assembly sequence. It continually monitors the PES and WMS and, in the event of an error situation being detected, suspends operation and attempts to diagnose and recover from that error. If automatic recovery is not possible, operator intervention may be requested. The three main components of the Error System, namely detection {error monitors}, diagnosis and recovery are illustrated in figure 3.

3.1 Plan Execution System

3.1.1 The Assembly Plan

The Plan Execution System {PES} [Duffy et al, 6] is responsible for working through the assembly sequence for the product and generating the required sequence of robot actions at run time. This assembly sequence is supplied by the user and consists of a set of instructions expressed in terms not unlike that which would be supplied to a human assembly worker. These instructions consist of primitive assembly operations and their associated parameters. The user specifies which robot should be used for which operations and uses preconditions for an operation's execution to ensure correct assembly sequencing.

The assembly sequence is loaded into a set of frames which form an assembly plan which are added to the Static Knowledge {figure 2}. These frames are worked on by the Job Manager to generate the correct sequence of primitive assembly operations which are added to the Task Pool. The set of primitive operations known by the system exist as a set of frames and their associated agents. These operations exist in a hierarchy with the least primitive operations being expressed in terms of more primitive ones. At the most primitive level the operations are expressed in terms of the universal command set. The user is free to express the assembly plan using any of the primitive operations.

All levels of the PES can interrogate the world model to obtain the dynamic information they require for their execution. However, the PES has no knowledge of the way in which the sensed information was derived and cannot directly instruct the world model to provide the information. If the required information is not available or is not of sufficient quality then the PES must wait until it does become available. If this does not occur within a preset time an error condition is flagged.

3.1.2 Universal Command Set

The universal command set is defined here to be the intersection of the sets of commands achievable by the individual robots within the work cell. These universal commands form the basis set from which all the primitive assembly operations can be constructed.

Each of the universal commands maps onto a single or sequence of actual robot commands. The number of corresponding commands and the order in which they are executed varies between propriety controllers. Associated with each robot controller is a robot interface agent. It is this agent's responsibility to map the universal commands issued for its robot into the corresponding proprietary commands and pass them, by some suitable communication protocol, to the controller for execution. In this way the peculiarities of the different robot command sets are hidden from the rest of the PES. In fact many of the features built into commercial robot controllers become redundant allowing future controllers to be much simplified.

3.2 The World Modeling System

The World Modeling System {WMS} is responsible for providing a description of the robots' working environment in a form easily useable by the PES and Error System. It must ensure that the information required by the rest of the system is available and is a consistent interpretation of a diversity of sensory inputs. For robust operation, the system must be able to utilise imprecise and incomplete sensory data [Durrant-Whyte, 7]. The main components of the WMS are the World Model {WM}, the World Model Manager {WMM}, the sensor input, and the sensor processing. These are illustrated in figure 4.

The World Model is the part of the WMS seen by the PES. It is a description of the working environment held as a set of frames in the WM pool. There are a number of levels of abstraction within the World Model. The highest level provides the best estimates of the positions of the objects within the workcell whilst at lower levels more specialised features are held, for example, a parametric description of the relationship between a robot's end-effector and a gripping point of an object. At the lowest level, the information may be sensor specific and is only useable by dedicated agents within the PES and Error System.

The WMM has three principle tasks. To combine the contributions from different sensor processing units, to maintain the consistency of the World Model under conflicting or deficient inputs, and to control the acquisition of sensory data to best support the operation of the PES. In order to achieve these tasks, the WMM draws on the Static Knowledge Base and information in the PES task pool as well as the processed sensory data.

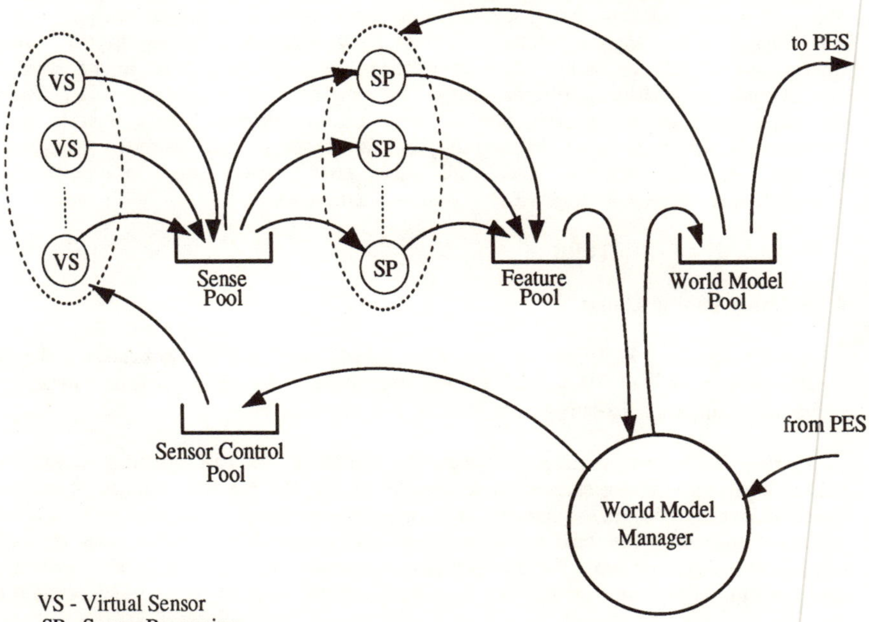

VS - Virtual Sensor
SP - Sensor Processing
PES - Plan Execution Sytstem

Fig. 4 World Modelling System structure.

Sensor input to the World Modeling System is presented as Virtual Sensor data. A Virtual Sensor is an abstraction of a physical sensor which hides the actual sensing mechanism from the rest of the system. This philosphy is similar to that of Henderson [8]. Virtual Sensor data is independent of the actual sensor from which it originated and so combination of data from different sensors is simplified. Virtual Sensor data may be further processed in isolation or in combination with data from other Virtual Sensors to provide higher level representations of the observed world. Control and guidance of physical sensors is also abstracted by the use of Virtual Sensors. A set of standard sensor control commands are used. These are mapped onto the actual physical sensor mechanism within the Virtual Sensor, again hiding the sensing mechanism from the rest of the system. The sensor processing uses the Static Knowledge Base and the latest information in the World Model in order to interpret the Virtual Sensor data held in the Sense Pool.

3.3 The Error System

The Error System continually monitors the operation of the system by means of a set of dedicated Error Monitors. These compare information held in the system's frames with internally generated expectancy, and, in the event of the observed information deviating beyond certain bounds, create a frame describing the nature of the error. The information in the frame includes the identification of the detecting agent, the time of detection and an error code. Other system agents may also create similar error frames if they detect errors in their own operation or data.

Error Diagnosis is triggered by the appearance of one or more of the error frames. The diagnostic system may seek further information if it cannot reach a satisfactory conclusion from the triggering frames. A satisfactory diagnosis is passed to the Recovery System which is responsible for formulating the recovery plan or, in the event of an unrecoverable or undiagnosible error, suspending operation. If a recovery is instigated, the current state of the assembly plan is saved and the recovery plan is substituted. Once the recovery plan is complete the original plan is restored and normal operation can continue. If the recovery plan fails, the error system may, in principle, operate recursively until recovery is achieved or operation is terminated.

4 PROCESSING STRUCTURE

4.1 Software Implementation

The present implementation is based around an IBM 6150 workstation connected by serial lines to a set of PCs and robot controllers. The PCs provide basic sensor processing facilities for the laser striper and the CCD video cameras.

The frame/agent structures run under the control of the AIX operating system on the 6150. Each agent is implemented as a separate process and the frames reside in an area of shared memory. The agents operate and interact concurrently under the multi-tasking operating system. Real time operation is not achieved on a uniprocessor such as the 6150 because of system overheads but the parallelism inherent in the software structure is designed to allow the system to be easily ported on to a parallel architecture.

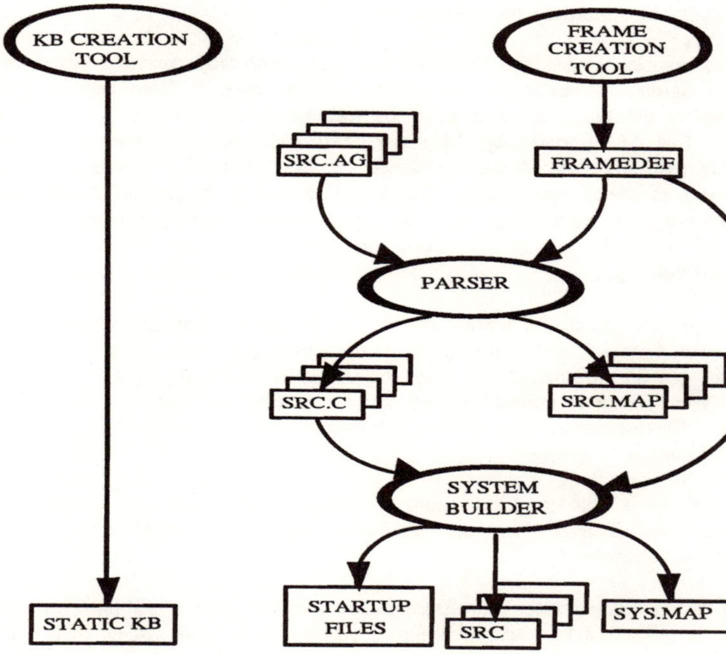

Fig 5. Overview of the Software Toolset.

4.2 Supporting Software Toolkit

A system of the type described in this paper will have two main categories of user. The production engineer, who is responsible for introducing new products and the development engineer, who wishes to improve the system's capability or performance. To provide a suitable interface to both these users, a set of software tools have been produced which are shown in block diagram form in fig 5.

The KB Creation Tool provides the interface for the production engineer. There are two types of static knowledge he must be easily able to impart to the system. The physical cell description - robot capabilities and configuration - is the first type and likely to change least often. The product description is the second type and consists of the part models and their assembly plan. Future developments may lead to this tool taking its input directly from a CAD system.

The work of the development engineer involves writing new agents, or modifying existing ones. The need for new agents will include the incorporation of new primitive assembly operations and their associated error monitors; robot interfaces for new controllers; and virtual sensor interfaces for new sensors. To allow the encoding of the new agent algorithms without the unnecessary details of frame implementation, the agents can be written in terms of high-level frame access functions which are mapped into the appropriate compilable code by the Parser. To create the frames associated with these agents the frame creation tool allows the exploitation of previously created frames and slots and prompts where necessary for required input.

With a complex system like that described, an automatically generated "map" of the parallel acting agents' interaction is valuable for system analysis. This map is generated in two stages. In the first stage a map of the frames/slots used by the individual agents is created. This is generated by the Parser as it transforms the frame access functions into compilable code. The second stage combines the maps of all the individual agents in the system into an overall system map. This is carried out by the System Builder which is also responsible for compiling all the agent code into executable modules.

5 CONCLUSIONS

This paper has described a control structure for an MRMS workcell which has been successfully used to integrate the operation of three robots with a variety of sensors. Assembly plans can be programmed at the object level without explicit reference to sensors. The sensory data is managed by an independent subsystem running concurrently with assembly plan. The use of sensor derived information during assembly greatly reduces the fixturing requirements.

The workcell is able to detect errors in its operation and, for a limited subset, automatically recover. If recovery is not possible operation is suspended and a request is made for operator intevention.

The current uniprocessor implementation does not achieve real-time performance. However, the design methodology adopted makes it possible to map the system onto a true parallel archictecture on which real-time performance may be achieved. The method of integrating the robot controllers, by embodying primitive assembly operations within the control structure, reduces the processing requiremts of the actual controllers. This means that robot controllers of the future will not need to support large and complicated programming environments.

6 ACKNOWLEDGEMENTS

The authors would like to thank the other members of the research team, SERC ACME directorate and IBM UK. Funding was provided jointly by SERC and IBM (UK) under SERC grant GR/D 69778.

7 REFERENCES

1. Henderson, T. C., Allen, P., Cox, I., Mitchie, A., Durrant-Whyte, H., Snyder, W., 1987, "Workshop on Multisensor Integration in Manufacturing Automation", Technical Report, Dept of Computer Science, University of Utah.

2. Koivo, A. J. and Bekey,G. A., 1988, "Report of the Workshop on Co-ordinated Multiple Robot Manipulators : Planning, Control and Applications", IEEE Journal of Robotics & Automation, vol 4, no 1, 91-93.

3. Kent, E. W. and Albus, J. S., 1984, "Servoed world models as Interfaces between robot control systems and sensory data", Robotica, vol 2, 17-25.

4. Minsky, M., 1975, "A Framework for Representing Knowledge", From: "The Psychology of Computer Vision", Ed. P.H. Winston, 211-277.

5. HMSO, 1983, "The Official Handbook of MASCOT"

6. Duffy, N.D., Herd, J.T. and Philip, G.P., 1988, "A Structured Approach to the Control of Parallel Acting Cooperating Robots". Proc 18th Int Symposium on Industrial Robots, 323-329.

7. Durrant-Whyte, H.F., 1987, "Consistent Integration and Propogation of Disparate Sensor Observations", International Journal of Robot Research, Vol 6, part 3, 3-24

8. Hanson, C., Henderson, T.C., Shilcrat, E., Fai, W.S., 1983, "Logical Sensor Specification". SPIE vol 449 Intelligent Robots: Third International Conference on Robot Vision and Sensory Controls, RoViSeC3 (November 1983).

Chapter 20

An analysis and simulation model for robot-based assembly systems

E. K. Lo and A. M. Dore

1. ROBOT-BASED ASSEMBLY

1.1. Assembly Systems

With the state-of-the-art in robotics (1), the single-station, multi-task General-purpose Assembly system (Robotic Assembly Cell) is only applicable to a family of similar products consisting of limited numbers of largely identical component parts. As the complexity and variability of the products increase, inflexible toolings and restricted robot reach would inevitably limit the versatility of the system. Technical and production requirements would ultimately exceed the capability of a single station and necessitate the use of multi-station systems. Here, two configurations are possible. A number of the robotic assembly cells can be "loosely" connected such that they can be operated in parallel either independently or in a coordinated manner (Flexible Assembly). Alternatively, each robot station can be assigned to a single task and be linked to other stations in a fixed, flow-line manner (Programmable Assembly).

Of these two systems, Programmable Assembly demands least system versatility but incurs high capital investment and hence is only suitable for complicated products required in high production volumes. Flexible Assembly is by far the most attractive concept since it can be designed to handle both simple and complex products at mid or low volumes. It is anticipated that with the widespread use of robots and sensory devices, the Flexible (and Robotic) Assembly System will become increasingly popular.

1.2. Economic Justification

Various attempts have been made to justify the use of robots in assembly. Abraham et al (2), and later Csakvary (3), of Westinghouse Electric have developed a quantitative and structured procedure for the selection of assembly systems based on two charts and a set of product and production parameters. Lynch (4) has constructed an economic model to study the general characteristics of dedicated, manual and programmable assembly systems by considering system configuration and utilization efficiencies in the evaluation of unit assembly cost. Boothroyd, Poli and Murch (5) have analysed the economic performance of various types of assembly systems including indexing, free-transfer, and hybrid mechanised systems, multi- and single- station operator systems, multistation systems with programmable workheads and various type of transfer equipment, a single-station two-arm robotic assembly system, and a universal assembly centre.

A study of these models has shown that they are developed mainly to predict the economic justification of robot-based assembly among others. They are not structured to be used as a detailed design and evaluation tool for robot-based assembly systems.

1.3. System Analysis and Simulation

The assembly process is characterised by its inborn variability which is inherited from variation in component parts, the assembly operations, and the equipment used. Such a feature exhibits itself clearly in a single-station, multi-task Robotic Assembly cell because all of these variable factors occur and interact with each other simultaneously, at a single workplace, but with different probabilities of occurrence. In order to evaluate the economic justification of installing a capital-intensive robot-based assembly system in a factory, individual factors such as product requirements, part quality and equipment capability and reliability must be examined, preferrably quantified, and the cell must be analysed and modelled in a structured and detailed manner, taking into account other physical constraints imposed by the assembly process.

With the support of the ACME Directorate of the SERC, an analysis and simulation model is being developed in the University of Salford to evaluate the economic viability of Robot-based Assembly Systems. The development of such a computer model will assist industry to study the techno-economic viability of installing different types of systems and, for each system, different types of assembly cells which may have various levels of programmability and adaptability. This would provide product engineers with quantitative information for selecting and redesigning suitable products for robotic assembly. It would also assist system engineers to select, evaluate and design appropriate equipment and toolings as well as to plan and schedule the assembly process.

2. MODEL METHODOLOGY

2.1. Input Data

When a company wishes to evaluate the economics of assembling certain products using robots, its system engineer often starts by defining the company's assembly requirements and examining existing and potential resources. He would collect related product, production and factory information essential to the company. This includes:

- detailed product and part specification (eg, drawing, model, product structure, precedence, part tolerance)
- production requirements (eg annual demand, batch size, number of styles, design changes)
- factory information (eg, labour rates, shifts per day, rate of return)

He would also review the resources both internally and externally available to the company by identifying their individual characteristics such as:

- accuracy and repeatability
- physical and mechanical limitations
- programmability and adaptability
- capital, tooling and training costs

2.2. Rough Plan

Based on the product and production requirements, and one or more arbitary unit production times (eg, pick and place), the system engineer would estimate the overall capacity required by the proposed assembly system. Here, the assembly sequence is defined by identifying possible relationships between the global product (ie, the entire system's end products) and potential local products (ie, the intermediate products of each robotic cell). A rough plan can be generated to determine:

- the number of virtual assembly cells required
- the local products to be processed by each of these cells

Rough planning is an iterative process which enables the system engineer to relate the product and production requirements to the capacity requirements. Each virtual cell is characterised only by an arbitary pick-and-place time and is not confined to any physical equipment to be used.

2.3. Detail Cell

The next step is to define the detailed configuration of each of the identified virtual cells. This would involve:

- the selection of appropriate equipment
- the layout of the cell
- the organisation of the assembly operations

Because of the kinematics of the robot, the design and layout of the cell should be studied in three-dimensions. This would enable the assembly environment be represented in a such way that detailed estimation of the theoretical assembly time is possible.

2.4. Simulate Cell Performance

In order to account for the tolerance specification of the parts, inconsistency in equipment performance, and variations due to other external causes during robotic assembly, the theoretical assembly time calculated based on the detailing of a cell must be further processed. This can be achieved by incorporating relevant probability of occurrence to the individual factors.

In the case of part presentation using a conventional vibratory feeder, one or more of the following attributes may contribute to the failure of the operation:

. empty feeder due to supply shortage
. feeder malfunctioning
. parts stick or tangle together
. parts out of tolerance
. parts contaminated by foreign parts and materials
. parts damaged or deformed in transit

A probability of occurrence can be attached to each of the identified attributes to determine realistically the chance that no part is available when the robot arm is due to acquire one.

Similarly, during the process of part acquisition and mating by the robot, probability factors can be assigned to the following parameters to assess the chance of unplanned collision or miss:

. part location and orientation
. part dimensions and tolerances
. robot and gripper positions
. robot and gripper accuracies and repeatabilities
. position, orientation and dimension of the mating parts

A realistic assembly cycle time can be determined for each assembly cell based on these and other factors representing the success rates of, say, tool changes and sensor detections.

2.5. Detail and Evaluate Overall System

With each of the robotic cells designed and evaluated, they can be integrated together to form a Flexible Assembly System for the company using various material and resource handling equipment. The system engineer would select appropriate equipment to transfer local products from one robotic cell to another in a manner acceptable to the way production is being organised. He has to decide on the level of buffer stock to be incorporated in the system in order to minimize the probability of total breakdown.

Finally, the output of the model would enable the system engineer to evaluate the average cost of assembling a part for all of the products identified by the company. It would also help him to study the economic effect of redesigning some of the component parts for assembly. It would provide a quantitative measure to justify the use of robots in assembly in a specific company environment.

3. INFORMATION MANAGEMENT

3.1. Database Management System

To store and organise the large amount of information for system analysis and evaluation, a CODASYL database management system known as D3M (6) is used to create, edit and retrieve requirement, resource as well as other related modelling data generated during the process of simulation.

Data is represented in a structure of records and links

organised in a network of relationships. A record can be common to two or more sets of relationships which it owns or is a member. The links between records are normally interpreted as physical pointers. Data is viewed and manipulated by the user as schema and sub-schema (ie, a subset of the schema the user wishes to view) using either a general purpose program or a routine-based data manipulation language.

3.2. Requirement Database

The Requirement Database represented by a sub-schema is part of a systems design database that handles data for a particular analysis and simulation exercise. It contains product and factory data as described in Section 2.1. which is specific to a company involved in the study. This data is manipulated using the routine-based data manipulation language which allows better system integration.

3.3. Resource Library

It contains data related to the various types of equipment used in a robot-based assembly system. This includes:

- robots (eg, Cartesian, SCARA)
- sensory devices (eg, tactile, visual)
- grippers (eg, dedicated, interchangeable, programmable)
- mating devices (eg, "floating table", active, passive)
- resource and material handling equipment (eg, feeders, conveyors, transfer lines)

Resource data is comprised of records associated with both the technical (eg, accuracy, repeatability) and economic (eg, capital, training and installation costs) aspects of a particular equipment. A link is provided to associate these with another record describing the geometry and perhaps, the kinematics of the equipment. Although this library is managed by a data manipulation language, it can also be viewed using a limited general purpose application program called D2V (6).

3.4. Model Database

Data generated throughout the execution of the analysis and simulation model is stored in this database which is represented by a sub-schema of a sytems design database. This data is often associated with modelling activities including:

- rough planning
- equipment selection
- assembly sequence determination
- cycle time estimation
- assembly parameter and probability calculation
- cost evaluation

4. MODELLING TECHNIQUES

4.1. Robot Kinematic Modelling

In order to detail a robotic assembly cell (Sec. 2.3.), the system engineer is required to model the layout and operations of the robot environment in three dimensions. This is carried out with the assistance of a computer software package called GRASP (Graphical Robot Application Simulation Package) (7).

GRASP is based on a solid modeller which creates robots and other equipment using simple primitives (eg, cuboid, prism and cylinder) and joint information (eg, type, position and acceleration limits). An application program can be developed interactively for a cell by defining the track of a robot arm relative to other equipment and toolings. The track program can be converted or 'dumped' as a process which computes the coordinate transformations of the joints and the theoretical cycle time of an operation. Animation, static and dynamic clash detections are also available.

4.2. Discrete-event Simulation

The stochastic, discrete-event simulation package HOCUS (Hand or Computer Universal Simulator) (8) is used to model the probablistic variable factors existing both in an assembly cell (Sec. 2.4.), and a flexible assembly system (Sec. 2.5.).

In discrete-event simulation, the state of a system of entities (eg, people or machines) at a particular time may alter as a result of an instantaneous occurence of an event which leads to changes in entity characteristics (ie, attributes). In HOCUS, an activity is a subset of a modelled event which allows appropriate features to be associated with entity attributes. This basic process involves taking an entity from a source queue, holding it for the duration of the activity, and then placing it in a destination queue. An activity will only occur if the correct entity is available from the source queue, and can be placed in the destination queue subject to all imposed constraints. Consequently, the probable occurence and duration of an event depends on the attributes of the related entities and the features of its activities.

4.3. Other System Functions

Apart from the above mentioned modelling techniques, the analysis and simulation system requires a third software module to process information which cannot be otherwise handled efficiently by either GRASP or HOCUS. This includes:

- low-level database manipulation (Sec. 3.2. and 3.4.)
- rough plan (Sec. 2.2.)
- equipment selection prior to cell layout (Sec. 2.3.) and cell integration (Sec. 2.5.)
- part presentation, acquisition and mating modelling

(Sec. 2.4.)
. cost evaluation (Sec. 2.5.)
. software module integration

These functions are performed by a software system called SARAS (Simulation and Analysis of Robot-based Assembly Systems) which is currenly being developed in PASCAL on a DOMAIN ring under a UNIX operating system. It enables the system engineer to model and evaluate an assembly system based on the proposed methodology using the GRASP and HOCUS techniques in an effective manner.

5. SYSTEM INTEGRATION

5.1. Software Architecture

The architecture of the analysis and simulation system is developed around three different operating environments: SARAS, GRASP and HOCUS. During simulation, the system engineer is working in one of these three modes depending on the activities being carrying out. Such an approach allows SARAS to be potentially independent of any third-party modelling and simulation packages. It also provides the engineer with the maximum amount of freedom and facilities when using the GRASP mode of operation. The three software modules communicate one with the other through ASCII text files.

SARAS is the overall controlling environment for the system. It ensures smooth transition to and from GRASP and HOCUS. Its user interface is written using APOLLO DIALOGUE routines and incoporates front-end features such as mouse-driven, icon representation, popup menu and screen picking.

5.2. SARAS/GRASP/SARAS Interfaces

The Resource Library described in Section 3.3. contains a library of records which describe the geometry of the component parts and the kinematics of assembly equipment in a text format acceptable to GRASP. Based on the choice of equipment for a cell, SARAS extracts appropriate records from the Library to form a GRASP input file. This is subsequently transferred to the GRASP environment automatically so that the user can define the layout and operations of the cells using the GRASP operation mode.

Simulation within GRASP can generate a text output file which contains process information such as:

```
: GRIP peg FROM bench TO robot AT 73.9806,
: GRIP peg FROM robot TO table AT 84.0991,
  .
  .
: MOVE robot
FROM 0.0000 TO 27.5944
```

Peg, bench robot and table are names of the objects defined in GRASP. The syntax "GRIP...FROM...TO...AT" represents the change of ownership of an object (ie, peg) which takes place at a particular clock time. The "MOVE...FROM...TO" statement shows the time taken to complete a robot movement. SARAS interprets this information. It indentifies and stores the following operation data:

. nature and sequence of the operation
. names of parts involved
. names of the equipment being used
. theoretical cycle time

5.3. SARAS/HOCUS/SARAS Interfaces

In HOCUS, a simulation model can normally be created interactively with graphic assistance, although the text input file that defines a model can be written directly. Typical modelling statements include:

```
ARUN
C  E  1MAN       1T          U1
```

The first character of a statement defines the nature of the data contained in that statement (eg, "A" defines an activity definition, "C" for an activity condition). The remaining characters may represent names, graphical data, or modelling operations depending on their relative positions. The first statement shown, for instance, defines an activity called RUN. The second statement describes a condition associated with the activity. This involves taking a previously-defined entity MAN from queue 1 (1MAN) by searching through the queue starting from the beginning (E) at the start of the RUN activity, and placing it at the tail of queue 1 (1T) when the activity is completed. During the activity, the function U1 is performed. This fixed-format text syntax allows SARAS to create a HOCUS input file when it is required, based on the overall modelling methodology and other information generated by, say, GRASP.

Data output from HOCUS can be obtained in an ASCII file. Consequently, SARAS can interpret the results for further processing.

6. PROGRESS TO-DATE

The research project was started in January, 1987 for three years with the support of ACME and three industrial companies: AMTeC, Jaguar Cars, and Rexel Engineering. Whilst AMTeC assists the project with advice on software development and commercial exploitation, Jaguar and Rexel provide company information to ensure the programme reflects realistically industrial situations. They have also supplied the project with products for model validation.

To date, the modelling methodology of the overall system and the robotic assembly cell has been defined. Software required for data input, rough planning, cell detailing and system integration has been implemented. An assembly cell based on a gantry robot is also being developed for two families of the Jaguar and Rexel products.

It is anticipated that the system will be completed for simulation and validation by June, 1989.

REFERENCES

1. Redford, AH and Lo, EK
"Chapter 2: The State-of-the-art in General-purpose Assembly"
ROBOTS IN ASSEMBLY
Open University Press (1986)

2. Abraham, RG et al
"Requirements Analysis and Justification of Intelligent Robots"
PROC. 5TH INTL SYMP. ON INDUSTRIAL ROBOTS
Chicago, USA (1975)

3. Csakvary, T
"Product Selection Procedure for Programmable Automation Assembly Techniques"
PROC. 2ND INTL. CONF. ON ASSEMBLY AUTOMATION
Brighton, UK (1981)

4. Lynch, PM
"An Economic Guideline for the Design of Programmable Assembly Machines"
ASME Publication 77-WA/Aut-2 (1977)

5. Boothroyd, G et al
"Chapter 7: Performance and Economics of Assembly Systems"
AUTOMATIC ASSEMBLY
Marcel Dekker Inc (1982)

6. Introduction To D3M
Order No. 001950
Revision 00
Apollo Computer Inc.

7. GRASP Reference Manual
B.Y.G Systems Ltd.
Nottingham

8. HOCUS User Manual
P.E Inbucon Ltd.
London

9. DOMAIN/Dialogue User's Guide
Order No. 004299
Revision 00
Apollo Computer Inc.

Chapter 21

Robot applications in the manufacture of aerospace composite structures

P. J. Armstrong and E. J. Wright

1. INTRODUCTION

A research project has recently been completed at Queen's University on the use of robots in the manufacture of aerospace composite structures. The work was sponsored by SERC and involved collaboration with Short Brothers plc. The latter provided manpower, materials and access to a Cincinnati Milacron T^3 robot for the duration of the project.

Composite structures are most often produced from a "prepreg" material supplied in the form of a thin woven fabric pre-impregnated with an epoxy resin. As shown in Fig. 1, pieces of the fabric or "plies" of varying size and shape are cut from the material and stacked or "laid up" in a moulding tool. Air is then removed from the mould and the assembled plies are subjected to a temperature and pressure curing cycle in an autoclave in order to produce a solid part. The part is then trimmed or routed to size and holes are drilled for subsequent fastening to other components. The integrity of the component is also tested, usually by scanning its entire surface area with an ultrasonic probe. Finally components are assembled to form structures, most commonly by installing rivets in the pre-drilled holes.

At present the production procedures described above are predominately manual. However, the demand for composite components and structures is increasing and the introduction of some form of automation is seen as essential in the longer term. Indeed it has been estimated that costs could be reduced by up to 50% through effective automation (1).

Some of the major aerospace companies have responded by developing special purpose automatic machines. However, hard automation has the disadvantage of high capital costs which can only be justified when significant numbers of identical or similar components are required. Such a situation is unusual in the aerospace industry, especially in the production of civilian aircraft, where batch sizes and indeed total production volumes are relatively small. Hence flexible automation involving the use of robots may be a more logical solution for many companies. The capital cost of an approach based on robots is likely to be considerably lower and robotic cells should be able to cater for a wide range of components, if the robot's programmability is fully exploited.

2. THE RESEARCH PROGRAMME

An initial analysis of composite manufacturing had identified potential robot applications in the lay-up process, ultrasonic inspection, drilling, routing and riveting. However, in each case an original robotic end effector would be required in order to fully explore the feasibility of the application. In addition, the experience of others in the aerospace industry had suggested that the conventional teach programming of robots would be unsatisfactory. Teach programming times had been found to be excessive and, if only a small batch of components is to be processed, the time required for teaching could exceed the time the robot was available for productive work. It was therefore decided that, if possible, off-line programming techniques would be adopted. Since such techniques are not as fully developed as the robot teaching approach, original research was required in order to develop practical off-line programming techniques.

A major disadvantage of off-line programming is that robot positioning depends on the robot's accuracy i.e. its ability to visit a numerically defined point in space, as distinct to its ability to repeat a taught point. Since robot accuracy is invariably inferior to robot repeatability, it was decided that sensors would be used in order to maintain errors in the position and orientation of the robot's end effector to within the necessary manufacturing tolerances. This policy coupled with the inherent complexity of the required end effectors dictated that the overall design of each end effector should incorporate a microcomputer for end effector control and the processing of sensor generated data.

In summary therefore the research programme involved a wide range of individual projects concerning end effector design, off-line programming procedures and the use of sensors for error detection and robot control. The intention was that each end effector would be mounted on the Cincinnati T^3 robot at Shorts to form, along with other purpose built equipment, an experimental cell. The cells would then be used to study the potential robot applications listed above. Prior to this, end effectors would be tested and developed at the University site and the GRASP robot simulation package would be used to pre-plan the experimental cells.

In addition a comprehensive computer network system installed at the university would be used to link a VAX computer running GRASP, two CAD systems, a robot cell supervisory computer and the microcomputers used for end effector control. This system would provide a facility for developing suitable software for the experimental cells and also for research on off-line programming techniques. The progress made in the case of each of the potential applications previously listed is discussed below.

3. INDIVIDUAL APPLICATIONS

3.1 Robotic Lay-up

In this case the operation to be performed by the robot involved the handling of composite material in the form of cut plies. The robot was to be capable of picking up individual plies of varying size and shape from different locations on a cutting table. The plies would then be stacked to form an assembly which would be cured to produce a flat component. Alternatively plies would be stacked in the correct sequence to form a kit for subsequent manual lay-up.

A vacuum gripper system was developed which is shown schematically in Fig. 2. The gripper surface is divided into 100 independent cells, each of which can be programmed to provide a vacuum or a positive air pressure if required. Information on the geometry and location of the ply to be lifted is used by the gripper's microcomputer control system in order to create the appropriate vacuum pattern. As a result a vacuum is created above the ply during pick-up and a positive air pressure is applied to adjacent plies so that they will remain in place.

The ability of the gripper to selectively pick and place plies was first tested using a purpose built pneumatic manipulator which was able to raise, traverse and lower the gripper head. The gripper was then mounted on the T^3 robot as part of an experimental lay-up cell.

In the cell a supervisory computer was used to program the robot off-line to pick plies from various known locations at one end of a table and stack them in the correct sequence at a fixed location at the other end of the table. A two camera computer vision system was also incorporated in the cell in order to compensate for robot inaccuracy, errors in the initial positions of the plies and the possibility of lateral slippage during ply acquisition. One camera was used to view the underside of the gripper after pick-up so that the position of the ply relative to the gripper could be checked. The second camera was mounted on the gripper and was used to locate a reference mark at the set down point. Information from the images obtained by both cameras was then used to compute appropriate adjustments to the position and orientation of the robot before the ply was released.

The experimental work carried out on the lay-up cell produced a variety of results and in general the gripper was found to be a satisfactory end effector for stacking plies on a flat surface (2). In this application the sensor approach used for error compensation proved to be successful and capable of reducing errors to within the necessary tolerances. The cycle times produced for the pick and place task were relatively long, but the vision system employed was particularly slow and a faster system would undoubtedly produce acceptable times.

3.2 Robotic Ultrasonic Inspection

The Company was particularly keen to automate the ultrasonic inspection of cured composite components, since the existing manual method of scanning the entire surface area of a large component was extremely time consuming. A joint investigation was therefore initiated whereby the Company would develop an appropriate end effector and the University would devise a suitable off-line robot programming technique, based on available CAD data describing component geometry. Following this initial work the Company proceeded to develop two robotic cells for the production scanning of composite components (3).

3.3 Robotic Drilling

In line with the overall policy of the research programme, the proposed automated drilling cell was to be designed to be suitable for off-line programming and equipped with sensors to detect any errors which may arise. The main requirement was that the robot should locate a drill at the correct position and orientation relative to a component prior to the drilling of a hole. Hence the possible sources of error concerned the relationship between the component and the drill. It was therefore logical to mount the necessary sensors on the end effector incorporating the drill. On this basis the end effector shown schematically in Fig. 3 was designed and built.

The end effector, which is referred to as a robotic drilling head, includes a self-feeding drill which is interfaced to a microcomputer and fitted with various sensors for drilling cycle control. The drilling head also incorporates a tripod of displacement transducers designed to make contact with the component (or a drilling template) when the drill is moved into position to drill a hole. The output from the displacement transducers is used to calculate the angular position of the drill relative to the component and the altitude or distance from the component to the drill.

A camera is mounted on the drilling head which views reference marks on the drilling template or the component fixture. The camera is connected to a vision system which is used to determine the position of the drill in the plane of the component. A second camera is also provided to check the critical distance from the drill to the edge of the panel. This camera is fitted with a fibre optic cable and lens assembly in order to provide a small field of view close to the component edge. Finally a load cell is incorporated in the nosepiece of the drill which measures the contact load between the nosepiece and the component. Physical contact was found to be necessary when drilling composite materials in order to provide support on both sides of the component. Without this support, hole quality was found to deteriorate. However, the contact load had to be controlled within appropriate limits and hence the load cell was introduced.

The outputs from all sensors incorporated in the drilling head are directed to the drilling head microcomputer where they are processed to produce signals suitable for correcting the robot's position and orientation.

Once all the necessary hardware and software had been developed, the drilling head was mounted on the T^3 robot and other equipment, including a purpose built drilling jig, was assembled to set up an experimental drilling cell. Thereafter a programme of tests was carried out in order to assess the drilling head and to investigate alternative drilling and programming techniques (4).

The test programme included tests on off-line programming whereby the processed sensor data was fed back to the robot controller on-line in order to reduce errors to within specified tolerances. However, sensory feedback was found to be only partially successful since on some occasions a large number of iterations was required before a sufficient reduction in the original errors occurred. This problem stems from the variable nature of the robot positioning errors produced when small moves are made close to the resolution of the machine. In turn the problem can be traced mainly to the approximate kinematic model used for co-ordinate transformation and any robot with revolute joints will exhibit this problem to a greater or lesser extent. However, it was not possible to test the sensory feedback approach with other robots which may be able to achieve satisfactory positioning without excessive iteration.

Since the main reason for investigating off-line programming was to reduce robot downtime to a minimum, an alternative approach to the latter problem was adopted, which would be suitable for a robot such as the T^3. This involved the use of teach programming but with the programmer presented with information from the drilling head sensors indicating the magnitude of the current position, orientation and contact load errors. The information was displayed on a monitor in a graphical form which indicated clearly the corrective action which the programmer should take. By using the sensor data in this manner it was found that the time required to teach the robot could be reduced by approximately 60 per cent. Furthermore the resulting robot program was more accurate than would have been possible if based on visual judgement. Hence it was concluded that sensor assisted teach programming as described above was the most suitable programming method for robotic drilling using the Cincinnati T^3.

3.4 Robotic Routing

Manual routing is performed using a template to guide the router as the component is trimmed to its final dimensions. Initial tests indicated that the robot would also require a template in order to produce the smooth routing path required. The robot programming task therefore became one of generating a sequence of points through which the router would pass, whilst at all times maintaining positive contact between the template follower (fitted to the router) and the template edge. Initially teach programming was used and a series of experiments was performed with an adapted hand router fitted to the T^3. The teaching process was found to be time consuming and the only method available to ensure a positive contact between the follower and the template was regarded as unsatisfactory.

This involved positioning the router so that the follower was just touching the template. The robot's teach pendant was then used to move the router a small distance along the inward normal to the template edge. However, this technique produced an indeterminate contact load which varied considerably from one programmed point to the next and also between programmed points.

As with the previous applications, off-line programming was to be used in the experimental routing cell. Despite the problems experienced with robotic drilling, the sensor approach to error compensation was again adopted, on the basis that robot positioning tolerances are much less stringent in template guided routing. As in the case of the robotic drilling head it was again logical to incorporate the necessary sensors in a suitably designed robotic routing head, which is shown schematically in Fig. 4.

The main feature of the routing head is the inclusion of a compliant coupling between the router body and the end of the robot arm. This consists of a spring loaded slideway which houses a displacement transducer in order to measure relative motion. By calibrating the coupling, the output from the displacement transducer can be used to provide a measure of the contact load between the follower and the template edge. Hence accurate control over contact load becomes possible as an alternative to the unsatisfactory procedure described above. Furthermore the presence of the compliant coupling means that the contact load becomes less sensitive to changes in robot position, which reduces the effect of robot inaccuracy and the variation in load between programmed points. For simplicity a unidirectional slideway was used which means that the robot must rotate the routing head as it progresses around the template (in order to maintain a normal relationship between the slideway and the template edge). Further calibration was also necessary to allow for the varying gravitational load on the coupling due to the weight of the router assembly.

Preliminary routing tests indicated that, apart from contact load, the only other variable which may require correction is the position of the follower measured along the axis of the router. A fibre optic rangefinder was therefore mounted on the front of the router body in order to measure the distance between the router and the template. The remaining sensors shown in Fig. 4 were included for research purposes and to compensate for the fact that the T^3 robot cannot output its current position and orientation whilst moving along a continuous path.

In the previous applications, sensors were used on-line to measure errors at programmed points. However, on-line sensory feedback is not possible whilst the T^3 robot is moving along a continuous path. Hence a two pass approach was adopted whereby the off-line generated program was first used to perform a trial pass with no router bit fitted. The outputs from the sensors were datalogged during this run, then processed and used to correct the original off-line programmed path. The intention was that the corrected program would then be run to rout each component in the current batch, with the sensors used, at this stage, merely to monitor errors retrospectively.

Experiments to assess the two pass off-line programming procedure demonstrated that it was a viable approach and that the robot downtime for the trial pass was not excessive compared to the overall floor to floor time for a typical batch. A further advantage was gained by introducing an automatic procedure to insert additional program points, if the variation in contact load exceeded specified limits during the trial pass. The advantage was that the original off-line program could consist of a sequence of relatively widely spaced points. Additional points would then be inserted where required as a result of the data generated during the trial pass.

3.5 Robotic Handling in a Riveting Cell

The proposed robot application in this case would involve a robot working in conjunction with an automatic riveting machine. The robot's task would be to acquire a temporarily fastened subassembly and to progress it through the riveting machine, whilst stopping at the appropriate points for holes to be drilled and rivets inserted. Since accurate positioning is required, it was apparent that the use of sensors to correct off-line programming errors would produce similar problems to those encountered with robotic drilling. Hence in designing a suitable end effector for the experimental riveting cell, provision was made for controlling the position of the component without using the robot. This led to the end effector shown schematically in Fig. 5.

As shown in the figure, the end effector uses suction cups to acquire the subassembly. The two stepper motors shown can be driven in opposite directions in order to adjust the distance between the suction cups to suit the geometry of different assemblies. However, by driving the motors in the same direction, lateral positioning of the assembly can be achieved. Hence independent positioning of the assembly in one dimension is possible, which for research purposes was regarded as sufficient to investigate the principle involved.

Apart from the gripper mounted on the T^3 robot, a table for incoming assemblies and a C frame incorporating a pneumatic clamp were included in the experimental riveting cell. The C frame was introduced to represent the automatic riveting machine, which in practice clamps the subassembly before drilling and riveting occurs. Hence from the robot's point of view, a simple clamping device and a programmed delay will suffice to simulate the operation of the riveting machine. Sensory feedback is provided by a camera linked to a vision system, which views the subassembly and registers any errors in its position before the simulated riveting cycle begins. For comparison the errors can be fed back to the robot or to the stepper motor control system in order to initiate corrective action.

The results obtained from the experimental cell demonstrated that the vision system coupled with the stepper motor drive system provides a fast method of correcting positioning errors in a single iteration. An effective off-line programming procedure was also demonstrated based on the use of a cell calibration technique to provide data for a GRASP simulation model of the cell. The GRASP model was then used to generate robot co-ordinates for inclusion in an off-line program.

4. CONCLUSIONS

The work reported above demonstrated that robots can be applied to a range of applications in composite manufacturing, providing suitable end effectors can be developed. The prototype end effectors produced for the research programme provided many detailed pointers for future end effector design,.

The viability of using multiple sensors was also demonstrated and in each case the sensor data produced was found to be consistent and reliable. Perhaps surprisingly alternative methods of employing the available sensor data evolved during the project, providing a range of options apart from the original concept of off-line programming coupled with on-line sensory feedback. This was in part due to the inaccuracy and limitations of the T^3 robot, but it also demonstrates a point not often stressed which is that there are several different approaches to robot off-line programming.

REFERENCES

1. Allison, H.B. and Pitman, A.A. "Future trends in technology and automation", Proc. AIAA International Meeting, 1980, Paper No 80-0922.
2. Wright, E.J., Curran, J.P. and Armstrong, P.J. "Robotic handling of composite materials in a lay-up cell", Proc. UK Research in Advanced Manufacture, Institution of Mechanical Engineers, London, 1986, pp39-46.
3. Campbell, N.A., Reid, I.M. and McClean, J.H. "Short's robotic ultrasonic scanning system", Proc UK Robotics Research, Institution of Mechanical Engineers, London, 1984, pp71-78.
4. Armstrong, P.J., Francey, S.D. and Wright, E.J. "Developments in robotic drilling", Proc 2nd National Conference on Production Research, Edinburgh, 1986, pp568-583.

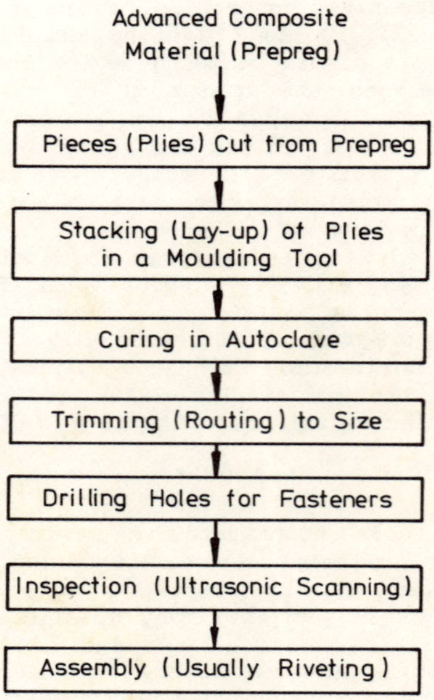

Fig. 1. COMPOSITE MANUFACTURING

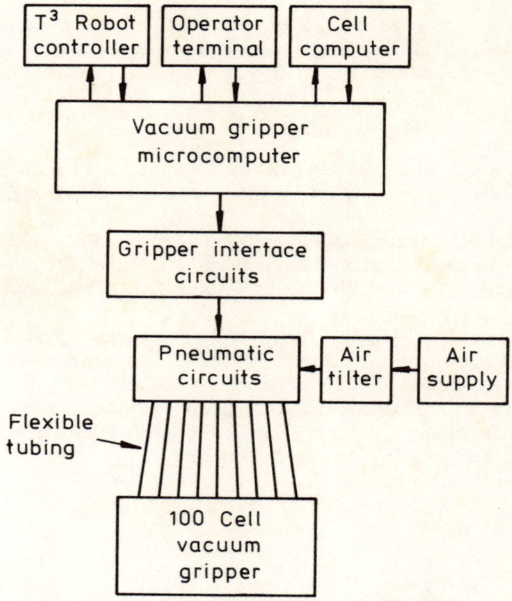

Fig. 2. VACUUM GRIPPER

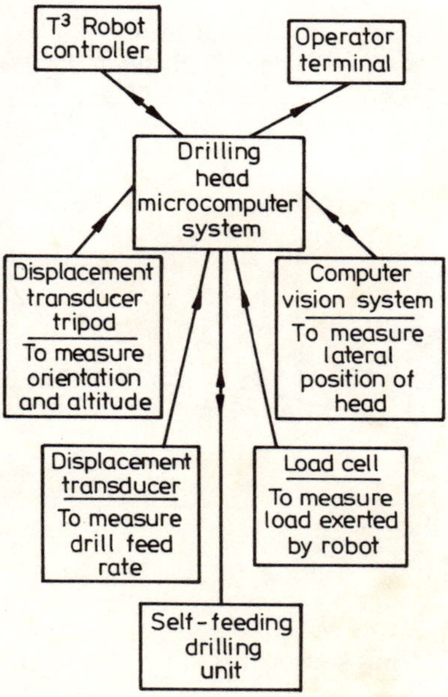

Fig. 3. DRILLING HEAD

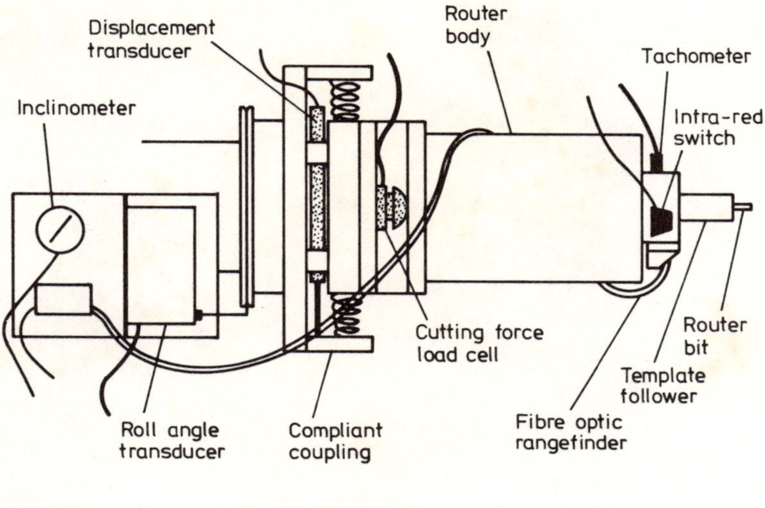

Fig. 4. ROUTING HEAD

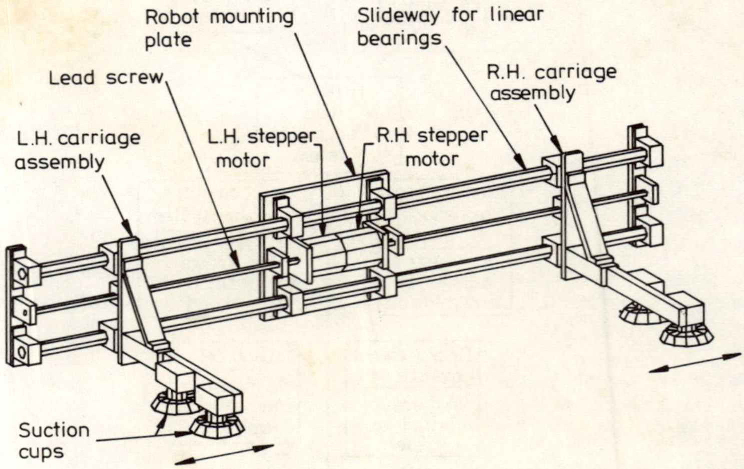

Fig. 5. RIVETING CELL GRIPPER

Index